GANN FOR THE ACTIVE TRADER

NEW METHODS FOR TODAY'S MARKETS

DANIEL T. FERRERA

COSMOLOGICAL ECONOMICS

WWW.COSMOECONOMICS.COM

COSMOLOGICAL ECONOMICS

THE MASTERS OF TECHNICAL ANALYSIS SERIES

The Masters of Technical Analysis Series brings together a collection of the most important classical and modern works on technical analysis and financial market forecasting. These classic works from the Golden Age of Technical Analysis were carefully selected by the late Dr. Jerome Baumring of the Investment Centre Bookstore in the 1980's, as representing the most valuable and important works in technical analysis ever written. They were included as the foundational source texts for his program in advanced financial market analysis and forecasting, and serve as the ideal foundation for any analyst seeking a thorough education in market theory and technical trading.

The Golden Age of technical analysis was a period from the early 1900's through the 1960's where the foundational theories of modern financial analysis were initially developed. The ideas and technologies developed during this fruitful period have formed the basis for most modern technical market theory, which is considered to be mostly a repetition or reworking of these past ideas and techniques developed by the Old Masters of the Golden Age. In these historical works can be found the timeless trading wisdom which has laid the foundation for all modern investment theory and literature. These techniques are as useful in today's markets as they were in the past, providing rare and valuable insights, tools and strategies that give the modern trader an edge over traders and investors that are unaware of these time honored tools.

Each quality reprint of these classical texts has been reproduced as an exact facsimile of the original text, maintaining the original layout, typeset, charts, and style of the author and time period, helping to preserve and communicate a sense of the feeling of the original work that a reproduction in modern format does not capture. Many of these rare works and courses were originally printed in only very small private editions or as correspondence courses, so that the originals were easily lost or destroyed over time. Our reproductions of these important source works have been printed on acid free paper and bound in a quality hardcover format that will compliment any trading library and help to preserve this important resource for generations to come.

The series is also currently being digitized and archived for permanent digital preservation by the Institute of Cosmological Economics, creating a searchable reference library of market wisdom accessible globally and available in new digital formats to keep the knowledge fresh and accessible through new devices and technology as we advance further into the information revolution. To see our full catalog of hardcover reprints, new publications, and digital editions please visit our website at www.CosmoEconomics.com.

TABLE OF CONTENTS

INTRODUCTION

In writing this book, I wanted to pass on the fact that trading is a profession, just like any other traditional profession and as such should be run with a strict set of business operation rules. This is without a doubt the most neglected aspect of trading and is the reason 80-90% of traders fail when it comes to successfully trading or speculating in the markets. Therefore, my first intent is to show that there is a science to running this type of business, the same as in any other traditional business. Second, I attempt to logically show that there is also a science to the task of *selectively* trading in the various markets, the knowledge of which reduces the level of risk in trading to a level comparable with all other lines of business. There are very few books on these subjects even though they are the most critical factors in prevailing in the markets. Without this knowledge, a trader is like a chicken running around without his head. He may keep his feet moving for a while, but his outlook is easily predictable. What I have to share is not a "system" by today's computerized definition of the word. I present the essential elements needed to succeed no matter what approach is used. Without these elements, failure will occur eventually. Yes, there is a systematic procedure that must be followed, but it is not a so-called trading system. I present methods and techniques that I have personally used to make money from the markets, but like any strategy or approach, it would eventually lead to failure without a strict set of business management rules.

DEDICATION

This work is dedicated to my loving wife, who has always
been by my side through good times and bad with truly unwavering
supportand belief in all my endeavors.

To my two young boys, that they may grow up to appreciate how
amazing nature really is and perhaps follow up on all of my work and
research.

To my father, who taught me the real value of persistence, perseverance,
and patience and to my mother, who stimulated my mind daily as a
young child with regular morning conversations over breakfast.

To my grandparents, who have provided me with many fond childhood
memories that I continue to enjoy regularly upon reflection.

To all the great market masters of the past, who inspired me to look for
something more in the wiggly lines of market data.

To the creator of life, who designed this amazing
universe in which we all live.

I would also like to send out a special thanks to TradeStation
Technologies for granting me permission to reprint charts from their market
software program.

Trading is a Profession

Trading is a profession and nothing less. Most professions such as medical, legal, psychological, pharmaceutical, technical, mechanical, engineering, etc., require several years of education, study and effort. Not to mention the money required for learning the necessary skills. Then it takes many more years of actual practice to develop the skills and gain the experience to become successful. It is ironic how many of these same professionals attempt to become successful traders without even the slightest knowledge of the markets or more importantly, money management principles. This occurs because there are no barriers to entry in this business. If you want to enter this profession, you just need a few extra dollars to open an account and that is it. However, if you wanted to become a doctor or a lawyer or enter some other profession, you would have to pay for an education, study hard, take tests, and obtain a license, etc., before you can even have the opportunity to find a job or start your own practice. This same principle applies to going into business yourself. Even if you wanted to simply open up a party store or a pizza shop, you would have many barriers to entry. This is not true with the trading business, which is another reason so many people fail.

A Word About My Writing Style

You paid good money to hear what I have to say, so make sure that you squeeze and absorb everything out of every word. It is not my style to ramble on or narrate a story. I tend to be concise and to the point. If I wrote it, there is a very good reason why it is on paper. I have so many books in my library that can all be summed up in just a few pages without leaving out any important details. If I want to read a story, I will purchase a novel, but if I want to learn something, the author had better start teaching the required information without too much unnecessary fluff. Conciseness is my style of disseminating information, so please follow my advice and read everything carefully as you go through the book. Reading each section several times is highly recommended. I am not one to write about a single technique and then provide one or two hundred pages of charts illustrating the technique just to fill pages in a book. I have found from teaching individuals, that it is much more effective if they learn the concept and find their own examples. This improves their recognition skills and actually tends to speed up the learning curve.

CHARTS

Charts are utilized because they reflect the underlying psychology of all the participants involved and provide a past record of important price levels. Although many traders today rely entirely on computer-generated charts from various programs, I feel that subscribing to a printed chart service proves to be worth much more than the annual cost involved. By having a subscription to a chart service, you will find many more opportunities in markets that you may have completely overlooked due to a lack of interest. The trading business is just like any other business; you do it to make money. Receiving a regular set of charts on the various markets once or twice a month will provide you with many different perspectives (daily, weekly, monthly) that may expose opportunities that are simply much too good to ignore. In addition, you gain a feel for how the various markets are playing off one another.

Hand charting is a dying art form due to computers and chart publications. Maintaining a set of hand drawn charts on your favorite markets is strongly recommended! It takes very little time each day (only a few minutes) to update these charts. By doing so, you become much more mentally connected to the market and will have little difficulty in remembering important prices. Hand charting makes you much sharper and more alert to upcoming opportunities. It also gives you a better feel for the market that can develop into a type of intuition, which cannot be obtained by any other means. I have even gone as far as drawing intra-day charts by hand. I suggest that you resurrect this lost art if you are truly serious about taking on the trading profession.

In summary, charts are used extensively because they are the best graphical representation of human logic and emotions. They show us visually the cumulative effect of all the participants involved. Charts also expose potential opportunities that might have otherwise escaped your attention. Hand charting connects you to the market more so than simply reading a chart. There is a certain level of expertise that can only be acquired from the manual process of drawing a chart. Markets are like people; each has its own personality, habits and nuances. Your hand drawn charts will become your encyclopedia of knowledge of the market's action. Charts are much more than a simple history of past prices. They reflect and depict the ideas, actions, reactions, and emotions of the traders commercial interests and speculators involved. If you want to write like someone in particular, just begin copying their work verbatim a few times, and you will develop a feel for their writing style. If you want to paint like someone in particular, simply trace their artwork several times and you will develop similar skills and again obtain a feel for their style. Drawing charts by hand accomplishes this same intuitive feeling for the markets. This is not a requirement for success, but it is something that you should do as a standard operational procedure. Just because there are more efficient methods to get the task done does not mean they are superior! Try hand-drawing charts of the markets you are interested in for at least 2 months and I think you will be more than convinced of their value.

HOW MUCH DO YOU NEED TO START CORRECTLY

This is a difficult question to answer because the amount varies based on the market you want to trade and the amount of risk you are willing to accept on each trade. Some markets are more leveraged than others and some markets tend to have more price movement than others. For most people, $10,000 is a really good number to start with. I have started with as little as $1,000 several times trading for friends or family members and have run profits up to $10,000 in a year, so it is possible to start with less. Personally, I feel that if you do not have at least $5,000 that you could just walk away from and not lose too much sleep over, then you should not begin your trading career in the futures market. I would suggest that you begin trading options or stocks if this is your situation. The goal of trading is much different than investing even though both approaches are just trying to make profits on the price change of some underlying asset. In my opinion, the goal of trading is to turn relatively small amounts of money into large amounts. You should NEVER trade with any amount of money that has become meaningful. Once this occurs, you should always start over with small amounts. By operating in this manner, you will never go out of business and you will have little difficulty sleeping at night. This is a lesson I learned more than once and I learned it the hard way over 10 years ago.

COMMODITY BASICS

Before you can begin trading in commodities or futures contracts, you must have a full understanding of the leverage involved. For example, the margin required to control one contract of wheat is currently around $1,200. The current quote for July wheat is 395-$^1/_2$ (Three dollars and 95'/2 -cents). The contract size is 5,000 bushels and trades in cents/bushel. This means that this contract is currently worth ($3.95-$^1/_2$ x 5000) $19,775.00. You are controlling an asset worth almost $20,000 for a deposit of only $1,200. This gives you a leverage ratio of 16.5 times your margin requirement. Commodities are less volatile than stocks. It is the leverage that gives the illusion of volatility. For example, if the price of wheat moved a dollar, this would be considered a huge price move for this market. However, if a stock moves a dollar, it is not a big deal. Understanding the leverage factor involved is something most commodity traders overlook. They tend to focus on the potential reward side of the leverage but neglect to factor in the down side. If you are going to start from the correct beginnings, then you must understand how leverage works. If you purchase a $200,000 home with a 10% down payment and mortgage the rest, you are using leverage. If the home goes up in value to $255,256.00 in the next 5 years, this would give you a profit of $55,256 dollars on your $20,000 investment assuming that you do not have to make any interest payments. Margin on commodities does not charge interest, which is the reason I am making this comparison. Now, if on the other hand, something was environmentally wrong with your home or location and all of a sudden the price dropped to $125,000.00 you would have a $75,000.00 loss because you still owe the mortgage balance of $180,000. This is exactly the type of leverage that exists in commodities. They are not as volatile as people claim; they are just highly leveraged. As another quick example, the popular S&P E-Mini contract typically requires a margin of $5,000. The contract is trading around 1150. At $50 x the index, this gives it a value of $57,500, more than 11 times the margin. Commodity trading is like using OPM (other people's money). You are sort of borrowing the difference between the actual value and the margin, just like the mortgage example. If you are correct, you benefit substantially. If you are wrong, you still owe the difference and can lose substantially as well. The Index on the following page should be very helpful in gaining an understanding of the leverage involved. For most commodities, you can simply divide the contract size by 100 to determine how much a full cent or full dollar move is leveraged. For example, the grains (corn, wheat, oats and soybeans) are all 5,000-bushel contracts and trade in cents/bushel. Divide 5,000 by 100 and you get $50. This means that each 1-cent change in price is worth +/- $50. Bean oil would be $600; copper would be $250, British pound $625, etc. Make sure that you fully understand how price changes work in each market before you ever attempt trading in them. A commodities broker can be very helpful in this regard. In the majority of cases, you do not need their advice, but if you do not know the basics, they can be a good source of free information.

Contract	Symbol	Exchange	Size	Trading Hours in CT	Minimum Tick
Grains					
Bean Oil	BO	CBOT	60,000 lbs	9:30 AM - 1:15 PM	1pt = $6
Corn	C	CBOT	5,000 bu	9:30 AM - 1:15 PM	1/4 ct =$12.50
Oats	O	CBOT	5,000 bu	9:30 AM - 1:15 PM	1/4 ct =$12.50
Soybeans	S	CBOT	5,000 bu	9:30 AM - 1:15 PM	1/4 ct =$12.50
Soymeal	SM	CBOT	100 tons	9:30 AM - 1:15 PM	10pts =$10
Wheat	W	CBOT	5,000 bu	9:30 AM - 1:15 PM	1/4 ct =$12.50
Metals					
Copper	HG	COMEX	25,000 lbs	7:10 AM - 12:00 PM	5pts = $12.50
Gold	GC	COMEX	100 Troy oz	7:20 AM - 12:30 PM	10cts =$10
Palladium	PA	COMEX	100 Troy oz	7:30 AM - 12:00 PM	5pts = $5
Platinum	PL	COMEX	50 Troy oz	7:20 AM - 12:05 PM	10cts =$5
Silver	SI	COMEX	5,000 Troy oz	7:25 AM - 12:25 PM	1/2ct = $25
Currencies					
Australian Dollar	AD	CME	$100,000 AD	7:20 AM - 2:00 PM	1pt = $10
British Pound	BP	CME	$62,500 BP	7:20 AM - 2:00 PM	2pts = $12.50
Canadian Dollar	CD	CME	$100,000 CD	7:20 AM - 2:00 PM	1pt = $10
US Dollar Index	DX	NYBOT	1,000 x Index	7:20 AM - 2:00 PM	1pt = $10
Euro Currency	EC	CME	$125,000 EC	7:20 AM - 2:00 PM	1pt =$12.50
Japanese Yen	JY	CME	$12,500,000 JY	7:20 AM - 2:00 PM	1pt =$12.50
Swiss Franc	SF	CME	$125,000 SF	7:20 AM - 2:00 PM	1pt =$12.50
Financials					
Dow Jones Indu	DJ	CBOT	$10 X DJIA	7:20 AM - 3:15 PM	1pt = $10
Eurodollar	ED	CME	100,000	7:20 AM - 2:00 PM	1/2pt = $12.50
NASDAQ 100	ND	CME	$100 x index	8:30 AM - 3:15 PM	5pts = $5
S&P500	SP	CME	$250 x index	8:30 AM - 3:15 PM	10pts = $25
E-Mini S&P500	ES	CME	$50 x index	8:30 AM - 3:15 PM	1/4pt = $12.50
30-Yr T-Bond	US	CBOT	$100,000	7:20 AM - 2:00 PM	1pt =$31.25
Energies					
Crude Oil	CL	NYMEX	1,000 Barrels	9:00 AM - 1:30 PM	1ct = $10
Heating Oil	HO	NYMEX	42,000 gallons	9:05 AM - 1:30 PM	1pt = $4.20
Natural Gas	NG	NYMEX	10,000 MMBTU's	9:00 AM - 1:30 PM	1pt = $10
Unleaded Gas	HU	NYMEX	42,000 gallons	9:05 AM - 1:30 PM	1pt = $4.20
Softs					
Cocoa	CO	CSCE	10 tonnes	7:00 AM - 10:50 AM	1pt = $10
Coffee	KC	CSCE	37,500 lbs	8:00 AM - 10:45 AM	5pts = $18.75
Cotton	CT	Cotton Exchange	50,000 lbs	11:15 AM - 2:00 PM	1pt = $5
FC Orange Juice	OJ	Cotton Exchange	15,000 lbs	11:30 AM - 1:45 PM	5pts = $7.50
Sugar	SB	CSCE	112,000 lbs	8:00 AM - 11:00 AM	1pt = $11.20
Meats					
Feeder Cattle	FC	CME	50,000 lbs	9:05 AM - 1:00 PM	2.5pts = $12.50
Lean Hogs	LH	CME	40,000 lbs	9:10 AM - 1:00 PM	2.5pts = $10
Live Cattle	LC	CME	40,000 lbs	9:05 AM - 1:00 PM	2.5pts = $11
Pork Bellies	PB	CME	40,000 lbs	9:10 AM - 1:00 PM	2.5pts = $12

RISK DISCLOSURE

Since we live in a society that has to warn people that fire logs are flammable and coffee is served hot, I must provide you with a risk disclosure statement. The purpose of this material is to educate and provide you with knowledge that I have discovered to be valuable in the financial markets. There is no guarantee that these market structures will continue to work in the future, i.e., "past performance is not indicative of future results." You should understand that there is considerable risk of loss in the stock markets, futures markets, and option markets. Hypothetical or simulated performance results have certain inherent limitations. Unlike an actual performance record, simulated results do not represent actual trading. Since the trades may not have been executed, the results may have under or overcompensated for the impact, if any, of certain market factors, such as a lack of liquidity. Simulated trading programs in general are also subject to the fact that they are designed with the benefit of hindsight. No representation is being made that any account will or is likely to achieve profits of losses similar to those shown. Neither I, nor anyone else involved in the production of this material, will be liable for any loss, damage or liability directly or indirectly caused by the usage of this material. The data used in this material is believed to be from reliable sources but cannot be guaranteed. The examples and projections contained in this work are not to be taken as "investment advice." Ultimately, you are responsible for all of your investment decisions. If you are unwilling to accept this responsibility, then you should not invest in the financial markets at all.

THE MOST NEGLECTED TRADING DISCIPLINE

Money management is the single most neglected and overlooked key in the trading business. In my experience, money management is by far the most essential element to actually making significant money from the markets! Everyone spends such an enormous amount of effort, trying to find ways to optimize entry into the markets, or develop systems, indicators and strategies, but very few ever spend time learning how to control their risk and actually manage it to their favor. Everyone wants to know how to find the exact top or the exact bottom of a price move and get in at the most opportune moment. This infectious bug has bitten me myself, but I have corrected the issue by making money management higher in priority than the actual selection process in my trading. You would be wise to do the same! Money management is absolutely crucial in ordinary everyday life, so why would you think trading or any other business or profession would operate under different criteria?

Professional gamblers like Bobby Singer, one of the most successful blackjack players in history, will tell you that money management is the single most important key to taking money away from the casinos. Singer made his fortune gambling at blackjack, which is a game where the odds are stacked significantly against the player. Stock, commodity and option traders are very lucky because they can actually stack the odds well into their favor. If strict money management allows someone to make a fortune in a game where the odds are against you, imagine what it can do for you in the stocks, commodities or options markets where you can stack the odds well into your favor! Can you think of a single business that does not have some form of cash management? At this point, we must now add the element of the markets into our business equation. To apply proper money management skills to each and every potential trade entry, we must understand all of the following criteria:

(a)   We must fully understand the advantage we are exploiting in the market and realize that no matter how good it is, it is not completely infallible. This means accepting the fact that the markets are always right regardless of your opinion.

(b)   Knowing ahead of time how much we are willing to risk on any single trade and using stop loss orders or options to specifically control that financial risk.

(c)   Knowing ahead of time the approximate target objective or profit potential in each trade, so that we may determine if the risk truly warrants the potential reward. If the potential dollar reward is not at least triple the amount of risk, then the trade is unacceptable regardless of how perfect it may look.

(d)   Being fully aware that the business of profitable trading is purely a numbers game based on exploiting the factors of (a), (b) and (c) above.

(e)   Using all the market clues and available chart information to expose advantages and strictly limit our risks on every opportunity.

(f)   Staying 100% consistent to our trading and analysis approach, removing all emotions and as much subjectivity as possible.

(g)   Taking action consistently and immediately on all market signals (entry, exit, profit targets, stop losses, etc.)

(h)   Accepting losing trades for what they are without regret because you did not risk an emotionally significant amount and you followed your rules mechanically without bias or any emotional subjectivity.

WHY 3 TO 1?

In baseball, if you can reach a batting average of 300, which means that you successfully make it to at least first base after hitting the ball into fair play 3 times out of 10, you are considered a very good hitter. A batting average of 400 is considered absolutely phenomenal and is nearly an unachievable goal. The last player to accomplish this was Ted Williams, of the Boston Red Sox, who hit an astounding .406 in the year 1941. This same analogy applies to trading successfully. You only have to be a 300 hitter to make money consistently if you always apply good money management to every attempt at bat. Sure, you are going to take losses; it is part of the game, just like striking out is part of baseball. However, to be a really good trader you are not required to be right all the time. I have proven this to my satisfaction time and time again. I personally know traders that are only correct two or three times out of 10 and consistently make money each and every year. How is this possible, you ask? By taking advantage of good risk to reward opportunities and applying strict money management to every attempt. If you make $3 for every correct trade and only lose $1 for every time you are wrong, your trading profession will be profitable even if you are only 30% accurate in your trading decisions. You would lose $7 on the wrong trades and gain $9 on the correct trades. The true secret of successful speculation can be found in money management rules. My book, *The Keys to Successful Speculation* goes into this subject with much more detail and illustrates methods that provide tremendous risk to reward opportunities. Good money management can even allow you to make money on random entries. It is truly ironic how many people find money management boring and making money exciting. I agree it is not the most pleasant topic to read about, but the results speak for themselves. Want to put some real excitement into your trading career? Spend your time mastering money management and your trading results will be catapulted to another level.

W.D. Gann's Most Important Money Management Rules

* Divide your capital into 10 equal parts and never risk more than 1/10th of your capital on any one trade. This applies to all remaining capital as well.

* Always use stop loss orders to protect your trade and place it immediately after entering a position.

* Never overtrade by taking large positions. This would violate your capital rule. Remember " safety first."

* Never let a profit run into a loss. When the market moves in your favor and you have a profit that is double the amount of risk you were willing to take, move your stop loss order so that you will have no loss of capital if hit.
* When in doubt, stay out or get out.

* Trade in active, liquid markets.

* Don't close your trades without a good reason. Follow up the position with stop loss orders to protect your accumulating profits according to the rules.

* Accumulate a surplus of capital. This rule is very important. After you have made a series of successful trades, put some money into a surplus account to be used only in emergency or in times of panic.

* Never average a losing position. This is one of the worst mistakes a trader can make.

* Never get out of the market just because you have lost patience or get into a market because you are anxious from waiting.

* Avoid taking small profits and large losses.

* Never cancel a stop loss order after you have placed it when entering a trade.

* Avoid trading too frequently, getting in and out too often.
* Be just as willing to sell short as you are to buy long.

* Never change your position without a good reason based on set rules.

* Avoid increasing your trading activity after a long period of success or a series of profitable trades.

Remember to accumulate a surplus and do not be tempted to increase your trading unit too quickly. Success can go to your head and has ruined many otherwise good traders.

I have a huge library of trading or market related material. Very few authors discuss the importance of money management. They may mention it, but provide nothing user friendly in terms of application. I have placed this topic ahead of the actual trading selection and/or general market information due to its importance and priority to being a successful trader. Every trader should read, *The Richest Man in Babylon,* which is available as a FREE download at http://www.the-richest-man-in-babylon.com/

UNDERSTANDING THE BASICS

The basic premise behind traditional technical analysis and chart reading is that all the relevant information concerning the present and future concerns or beliefs about the specific market is reflected in the price chart. Therefore, all information, no matter how elusive or secretive, must first surface in the actual pattern of buying and selling or lack of, which moves the price of the market, providing clues to the future direction of the price movement. Traditional technical analysis also subscribes to the theory of a price manipulation paradox. The belief is that the markets are often taken higher in order to create a buying mania to provide a safe selling opportunity for the more powerful market forces known as "the manipulators." The market is often driven downward in order to stimulate a selling panic that will provide a safe buying opportunity for the more powerful market forces. It is believed that the " manipulators" use and rely upon the emotional response of fear and greed to motivate the weak-minded masses to buy or sell, thus creating opportunities for them since they are always on the other side of the coin.

I personally do not subscribe to this theory. I believe that the majority of traders do not have money management skills and thus they always react incorrectly when the emotional burden of a financial loss becomes too great to withstand. Fear and greed are still responsible for the entry or exit of the position, but stupidity or lack of knowledge is the underlying cause and not some powerful outside group of manipulators.

Markets are also seen as running in a cycle of four phases, especially stocks.

(1) ACCUMULATION - This is typically a quite low-level period, similar to flat lining on a heart monitor. Stock is slowly being accumulated "under the radar" from weak shareholders by those in the know. Volume is quite low and it is more of a "wearing out" process causing shareholders to lose patience, weakening their resolve to hold on to the stock another day. Markup

(2) MARKUP- Once the weak shareholders have been cast aside the stock is now in strong hands (people that will not sell for quite some time) and the stock begins to rise.

(3) DISTRIBUTION - Stock prices have neared or reached their peak price level, volume is high and the public is insanely excited into buying at a pace that far exceeds the rate at which they sell. This is when the stock is transferred from strong to weak hands. Distribution is comparable to cardiac arrest on the heart monitor. It's usually somewhat range bound, but the fluctuations are large and fast.

(4) MARKDOWN - The stock has been completely distributed to the weak hands of the public majority and the price begins to inevitably fall as additional buying dries up and the decline in price stimulates fear.

The chart on the next page graphically illustrates this basic four-phase cycle. The book *How to Make the Stock Market Make Money for You* by Ted Warren is an excellent reference source for a deeper understanding of this tune of analysis.

THE FOUR BASIC PHASES OR CYCLES OF STOCK PRICES

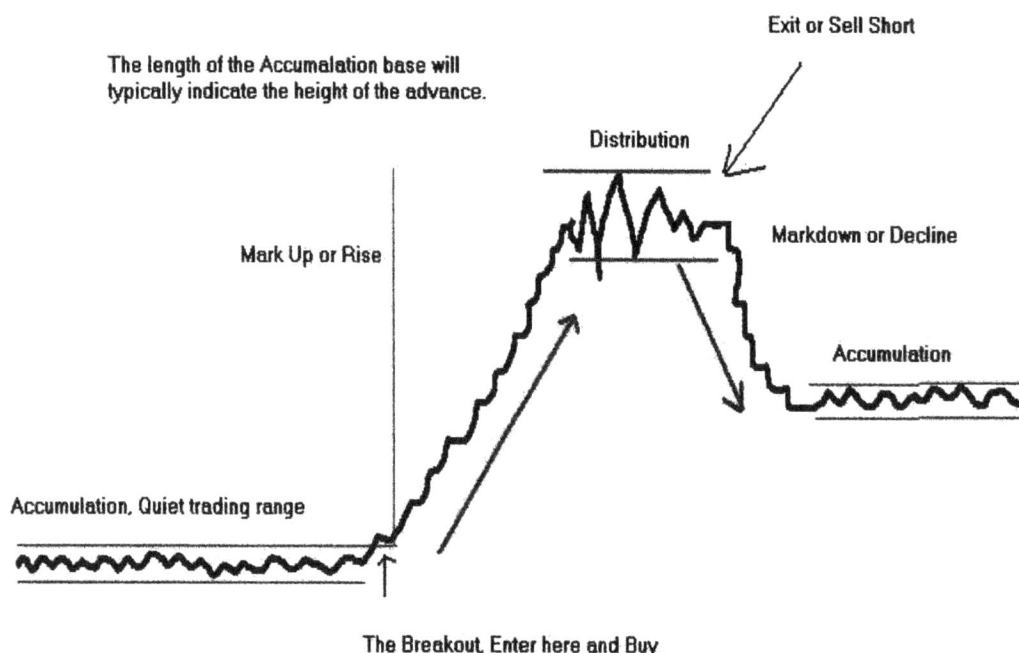

The length of the Accumalation base will typically indicate the height of the advance.

Exit or Sell Short

Distribution

Mark Up or Rise

Markdown or Decline

Accumulation

Accumulation, Quiet trading range

The Breakout, Enter here and Buy

There are several other chart patterns that are recognized for having predictive qualities. This book is not intended to provide information that is easily available from other sources, but it is necessary that you at least have an understanding of the basics and some of the underlying psychology that is involved in the interpretation of the chart pattern.

THE HEAD & SHOULDERS TOP AND BOTTOM FORMATION is probably one of the most highly recognized chart patterns in traditional technical analysis. The depressing influence of a long term head & shoulders bottom pattern on a monthly chart has a good track record in preceding a significant rise in a stock or commodity market. Again, this must not be a short-term pattern. Most take at least a full year or more to form. Typically, there has already been a significant downtrend that precedes the pattern. Then the market quiets and begins to consolidate like an accumulation pattern. Then the public is frightened out of their position during the first shoulder and head of the pattern, known as a "shakeout." During the second shoulder, they sell out their positions in fear that the price will drop again. They have lost hope in the market's ability to rise, which is exactly the reason it does. The rule with both head & shoulder patterns is that the longer it takes to form, the more reliable it is. The head & shoulders top formation suckers the public in with a rapid explosive type advance. This is a frequent distribution pattern in a large variety of markets. Once the neckline is broken, the market will typically move the same amount of points that measures the distance the head is from the neckline.

22

For example, if we have a "head & shoulders top" and the head is 35 points from the neckline, the market would be expected to decline 35 points once the neckline was broken. The inverse is true for the "head & shoulders bottom." The chart below illustrates an ideal pattern for both.

There are of course other traditional technical analysis patterns that show up in price charts with much greater frequency, especially double tops and double bottoms as well as triple tops and triple bottoms. The chart below illustrates these fundamental market patterns.

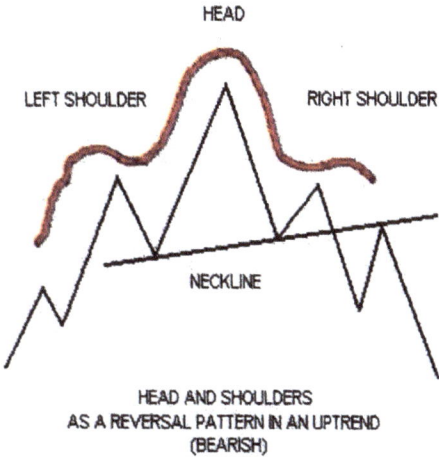

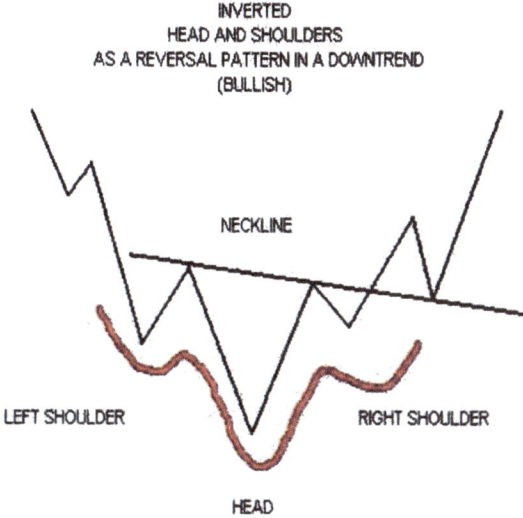

HEAD

LEFT SHOULDER   RIGHT SHOULDER

NECKLINE

HEAD AND SHOULDERS
AS A REVERSAL PATTERN IN AN UPTREND
(BEARISH)

INVERTED
HEAD AND SHOULDERS
AS A REVERSAL PATTERN IN A DOWNTREND
(BULLISH)

NECKLINE

LEFT SHOULDER   RIGHT SHOULDER

HEAD

Double Tops

Triple Top

Double Bottoms

Triple Bottom

The above patterns repeat frequently in all markets and provide very good risk to reward ratio. Recognizing these simple chart patterns can give you a simple but effective advantage over the markets.

UNDERSTANDING TRENDS AND TREND LINE BREAKS

One of the primary reasons for studying a chart of the market is to determine the trend or direction of the price movement. As traders, it is often beneficial to follow the path of least resistance. At least until we have a solid topping or bottoming formation such as the ones already illustrated on the prior pages. Up trends are a series of higher lows and higher highs. Up trends can also be defined as a series of advancing swings with higher tops and higher bottoms. Down trends on the other hand are a series of lower highs and lower lows. Down trends can also be defined as a series of declining swings with lower highs and lower lows. The charts below provide examples of both definitions.

Up trends on bar charts have a series of higher highs and higher lows.

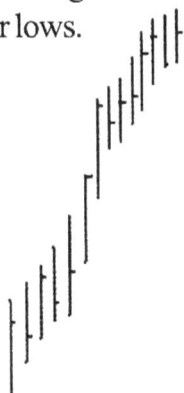

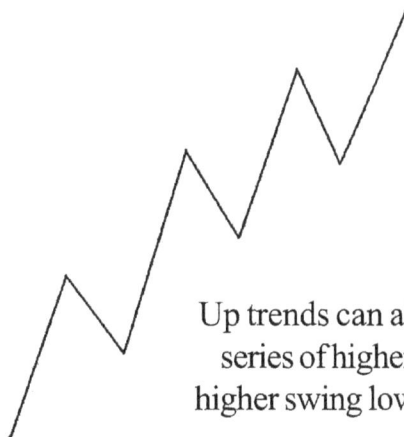

Up trends can also be defined as a series of higher swing highs and higher swing lows on a swing chart.

Down trends are a series of lower lows and lower highs on a bar chart. On a swing chart, we would see a series of lower swing highs and lower swing lows.

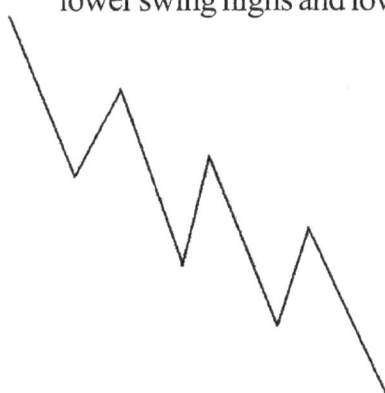

As you can see from these illustrations, the "bullish" pattern or rising trend has the basic requirement of rising swing lows. In its most basic form, each individual bar would have a higher bottom on each successive bar and over several days or weeks; each corrective swing would also end up at a higher swing low than the previous swing low. The term "swing" will be made much clearer as we progress through these basics. For now, just understand that we are simply talking about a price move up to a high point, and then back to a low point, then back up again to a new higher point.

Trend Lines are a connection of at least two swing highs or two swing lows that approximates the slope of the market's rate of advance or decline. By definition, only one line can be drawn between any two mathematical points. The two points are not the line, but they represent a segment of the line and also the portion of the line that exists between the two given points. In the markets, a good trend line should connect at least three successively higher swing lows for an uptrend and three successively lower swing tops for a downtrend. The next chart illustrates this requirement for a valid trend line.

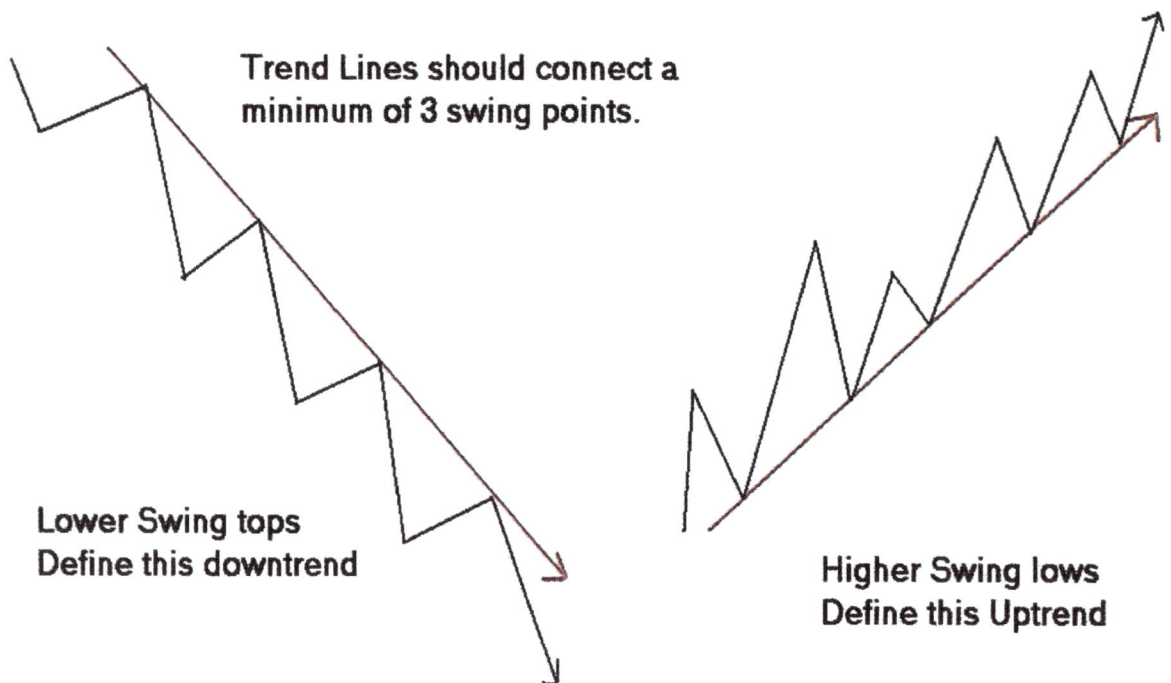

Trend Lines should connect a minimum of 3 swing points.

Lower Swing tops Define this downtrend

Higher Swing lows Define this Uptrend

26

MARKET SWINGS

Swings are somewhat more difficult to define because there are many methods to measure them and or filter them. Basically a swing is a directional movement that connects a low point in price to a higher point in price back to a lower point in price and so on. Each transition from the low to the high point, no matter how small or large the scale, is referred to as a swing. In mathematics, we would define this as a vector. Vectors are line segments that have a specific direction as well as defined starting and ending points. To make this clearer, we will look at the following examples.

S&P500 Index

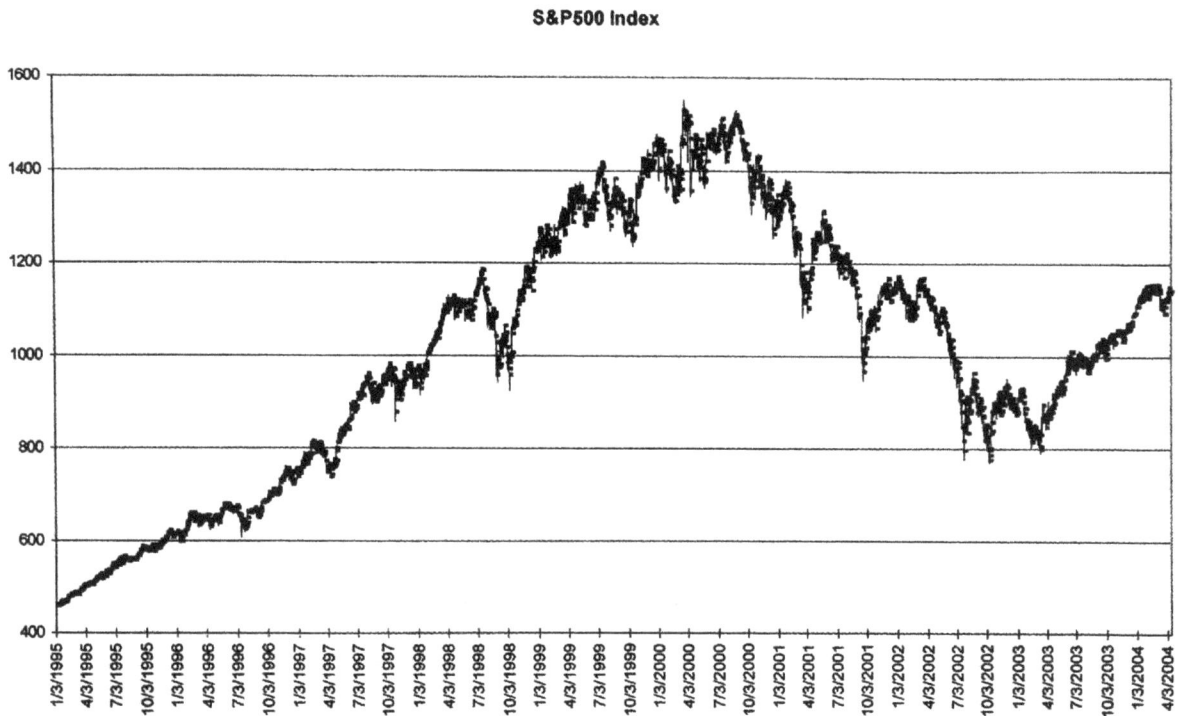

Looking at the above chart, we could simplify nearly 10 years of this market's movement utilizing three distinct market swings or vectors. The chart on the following page shows us exactly how this would look.

S&P500 Index

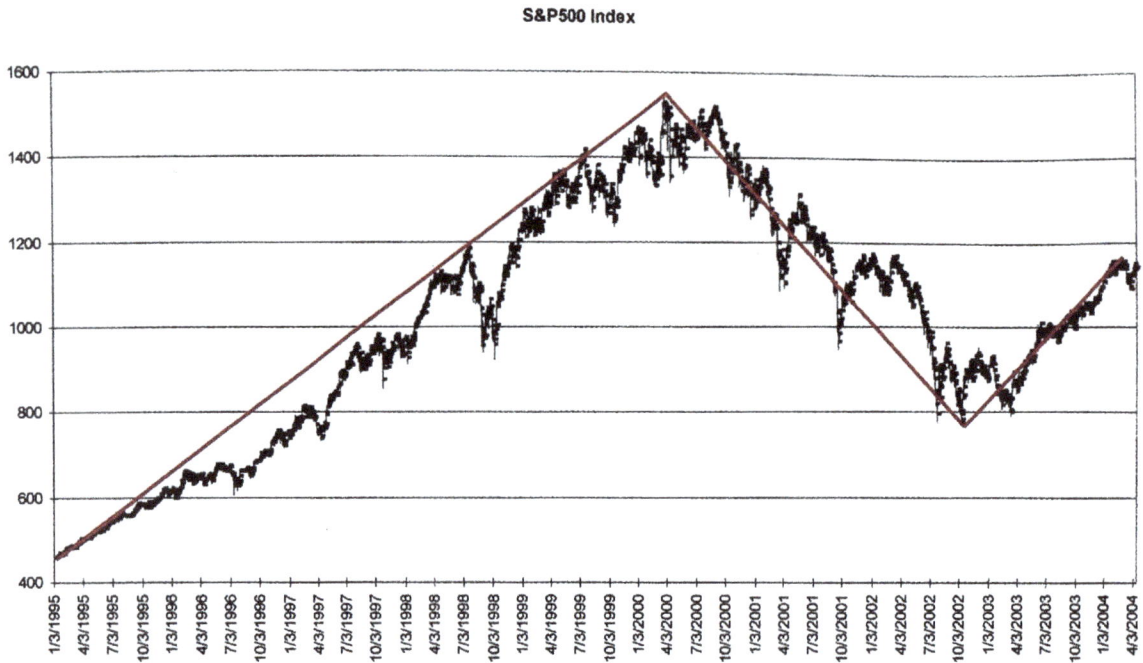

Here, 3-vectors (up, down, up) are used resulting in a much less complicated presentation of the data representing the exact same stock market curve. Obviously, we can increase or decrease the level of resolution by changing our definition of the minimum swing size. This could be in terms of a percentage change in price, or it could be some fixed minimum size in terms of points. The next chart provides an example of this concept.

S&P500 Index

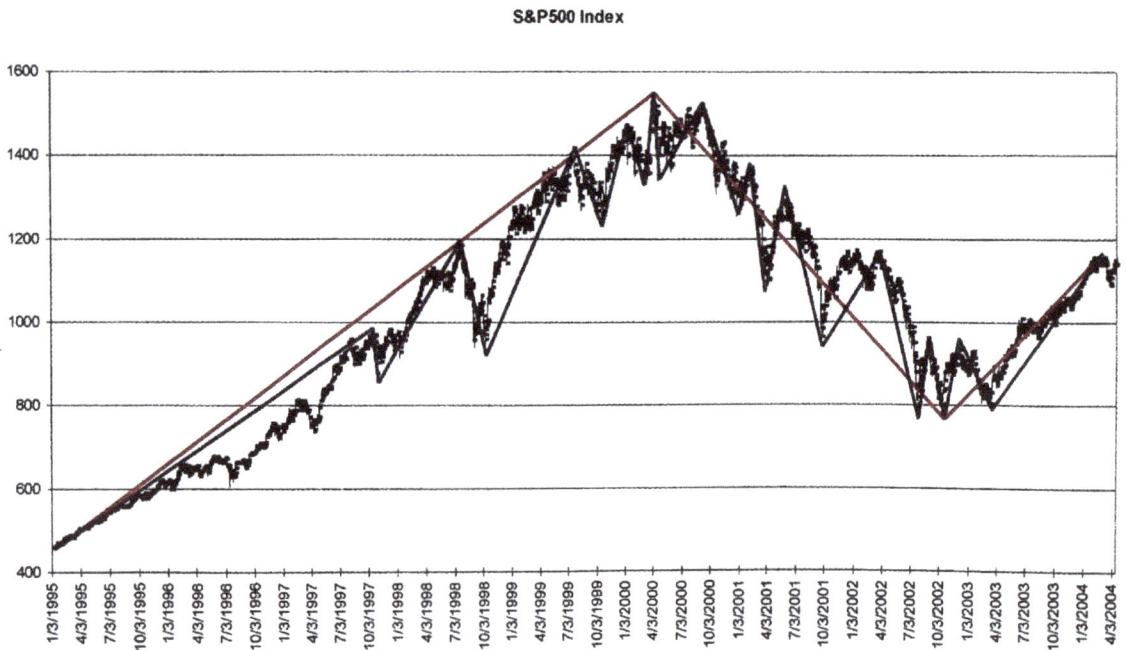

In the last chart, we have added several more swings to increase the resolution, but it still simplifies the data significantly. Now, we can see the markets in another light. Defining swings in a market has many useful functions, which will be described in more detail as we progress further into the material. Needless to say, I find this to be one of the most practical approaches to viewing any market on any time frame. In the markets, swings represent the terminal points or prices where a significant

28

number of decisions were made that changed the direction of the price movement. In physics and nonlinear mathematics, this phenomenon is known as *"Instantaneous Balanced Stability."* This condition occurs when all potential energy reaches a minimum or a maximum. In the markets, these *" decision points"* or price areas define boundaries of support and resistance levels where future key decisions are likely to occur again. Swing charts also make it much easier to identify or at least bring to your attention the classical chart patters we reviewed earlier. If you quickly review the chart patterns on pages 23 through 25 and then look at the swing chart again you will see what I mean. We can add or remove as many swings as we personally like relative to our trading style or approach. As mentioned earlier, swings can be based on a minimum amount of price change or it could be based on a percentage. This will be discussed again later on in the material.

SUPPORT AND RESISTANCE

Support is basically a price area where a declining market is expected to stop. Many people use the analogy of a floor. If you drop a ball, it will stop falling when it hits the floor and will bounce back up. Support can be a horizontal price level based on previous market bottoms or it can be diagonal based on the projection of a trend line connecting a series of higher lows.

Resistance is a price area where a rising market is expected to stop. The analogy of a ceiling is often used for describing resistance. If our bouncing ball rises high enough to hit the ceiling, it will stop and begin to fall back towards the floor. Resistance can also be horizontal based on prior market tops or it can be diagonal based on a trend line connecting the series of lower tops. The chart on page 25 illustrates diagonal support and resistance.

It is important to note that the price action of various markets will often consolidate into narrow trading ranges around support or resistance levels.

Support & Resistance can be Horizontal or Diagonal

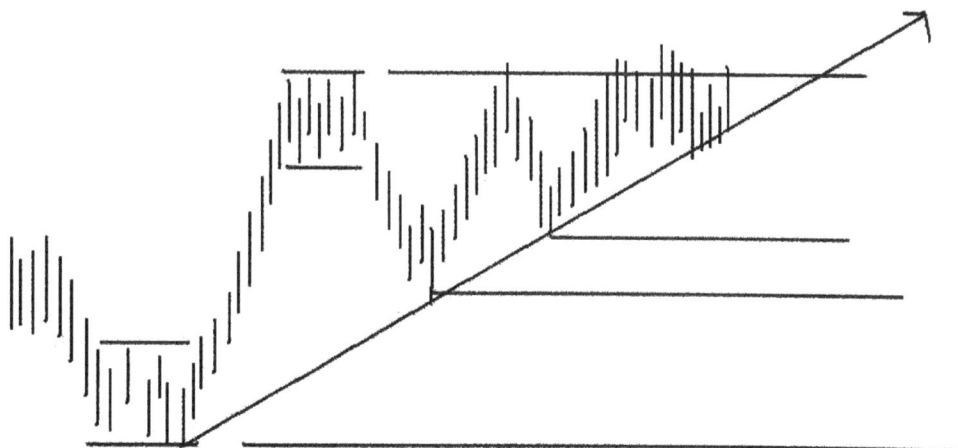

Looking at the simple illustration on the previous page, we can see the diagonal support on the rising trend line as well as the horizontal resistance at the highs. We would also anticipate that if the trend line were broken to the downside, each low area that acted as support previously during the advance, will likely act like support again as the market declines. Much can be learned about the behavior of markets from simply carrying these horizontal price levels across on your charts.

SHORT TERM CONSOLIDATION PATTERNS

When a market gets into a tight trading range for a minimum of 4 to 5 days, it provides a good risk to reward opportunity for the astute trader. The risk involved with this type of trade setup is limited to the height of this narrow range. The majority of the time, once a market breaks out of the narrow range, it will continue in the direction of the breakout until it reaches the next support or resistance point on the chart. If you have been diligent in carrying across your horizontal support and resistance prices, you will know exactly where the market is anticipated to stop. Knowing this information will allow you to determine the likely profit target so that you may objectively compare the potential reward to the amount of risk involved. The chart below provides an example of what these setups tend to look like.

4 to 5-day consolidation patterns

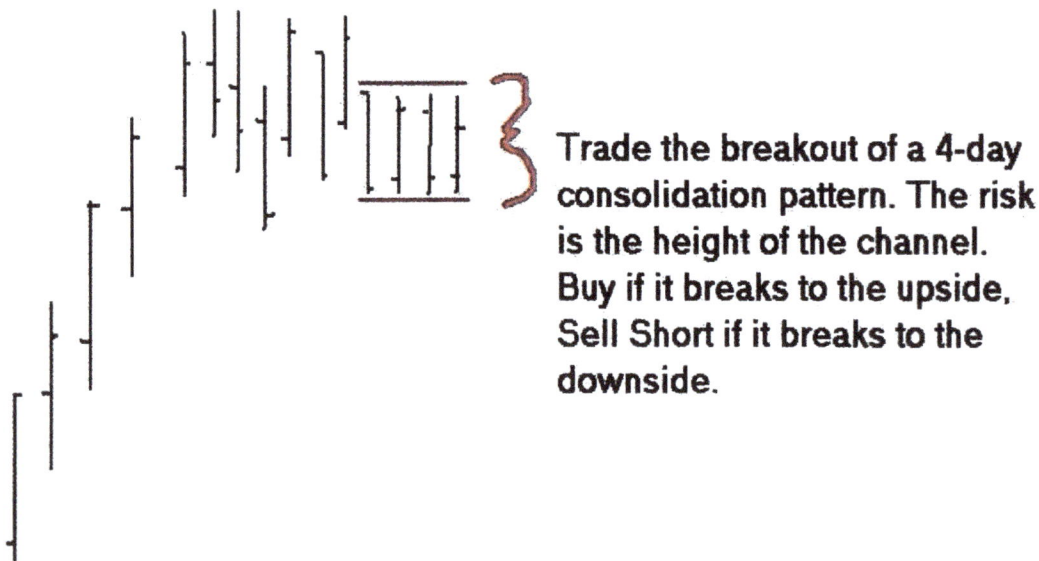

Trade the breakout of a 4-day consolidation pattern. The risk is the height of the channel. Buy if it breaks to the upside. Sell Short if it breaks to the downside.

TRENDS AGAIN
Bar Grouping Technique

Now that you have a better understanding of the definition of a trend, it is important to view them with more than one perspective. Short-term trends may be seen on an hourly or daily chart. Intermediate trends may manifest more clearly on a weekly chart, while long-term trends show up nicely on the monthly charts. Having more than a single perspective of trends, as well as support & resistance, provides extremely valuable information for making trading decisions. For example, if you are trading based on the daily chart, the trend lines on the weekly and monthly charts can be very useful in planning your entries. For example, if the weekly trend is UP and the market is experiencing a short-term decline on the daily chart, selling short is probably not the best strategy. You would be much better off to look for potential support areas to enter long positions in the direction of the bigger trends. On my hand drawn charts, I will often use a technique called "bar grouping," where I draw the larger trading bar around the smaller. This way I can easily see the smaller trends that are riding on the larger trends. The technique is simple and only requires you to define how many bars are in each group. For example, you could make a 3-day group if you wanted to see a daily chart inside a 3-day chart. You could make a 5-day group to see the daily inside the weekly or a 20-day group to see a daily inside a monthly, etc. The following example should illustrate the method clearly. For this example, I used TradeStation's rectangle drawing tool to manually group each set of 5-bars together, allowing us to see the weekly chart and the daily chart together. The boxes are drawn based on the high and low for the week creating another bar chart.

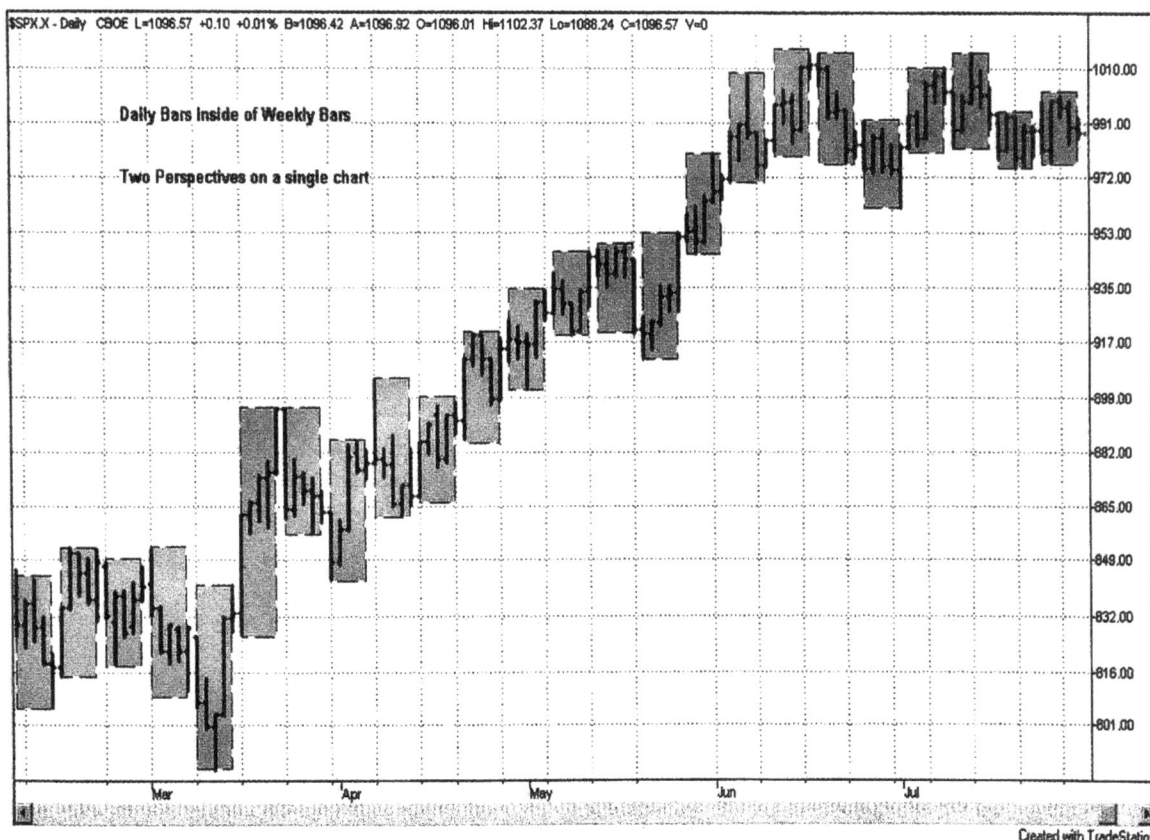

Daily Bars Inside of Weekly Bars

Two Perspectives on a single chart

Created with TradeStation

Bar grouping is a simple but very useful technique that provides a multi-perspective view to your charting analysis. For example, would you have spotted the 4 to 5 bar consolidation pattern without this view?

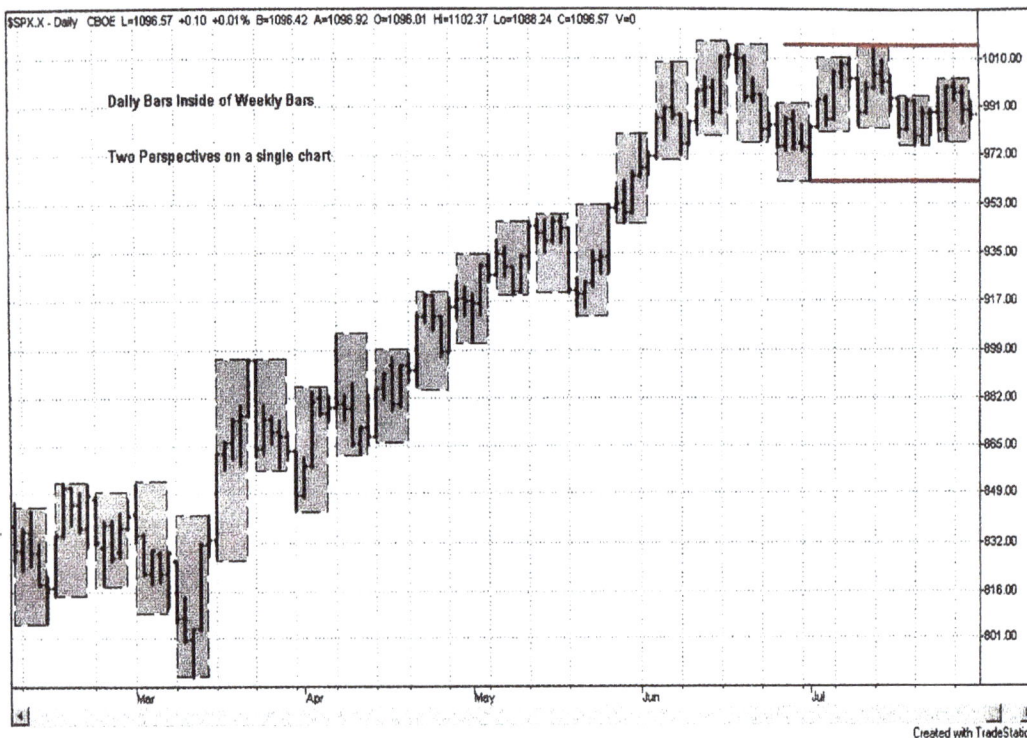

$SPX.X - Daily CBOE L=1096.57 +0.10 +0.01% B=1096.42 A=1096.92 O=1096.01 HI=1102.37 Lo=1088.24 C=1096.57 V=0

Daily Bars Inside of Weekly Bars

Two Perspectives on a single chart

Created with TradeStation

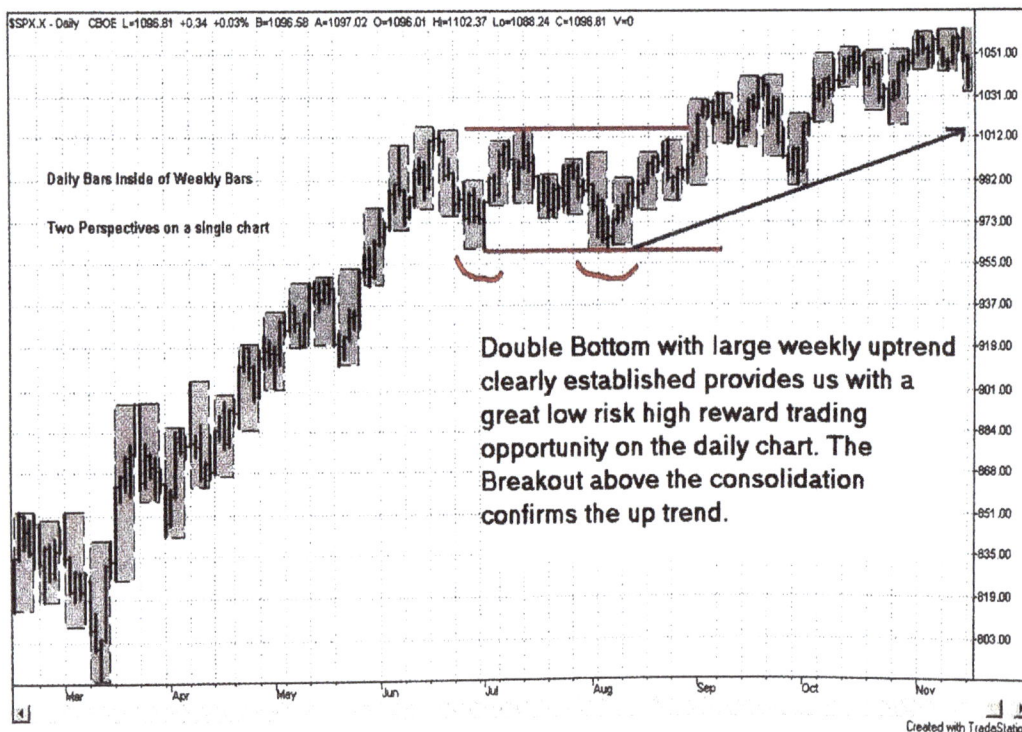

$SPX.X - Daily CBOE L=1096.81 +0.34 +0.03% B=1096.58 A=1097.02 O=1096.01 HI=1102.37 Lo=1088.24 C=1096.81 V=0

Daily Bars Inside of Weekly Bars

Two Perspectives on a single chart

Double Bottom with large weekly uptrend clearly established provides us with a great low risk high reward trading opportunity on the daily chart. The Breakout above the consolidation confirms the up trend.

Created with TradeStation

The primary usefulness of trend information is to prepare and develop entry and exit strategies for trading. If we are generally "bullish," we wait for buying opportunities such as double or triple bottoms or a price retracement that finds support on our daily, weekly or monthly trend lines. If we are "bearish" and anticipate lower prices, we sell short at each double top or triple top as well as price retracements that find resistance on our daily, weekly or monthly trend lines. Bar grouping allows us to see more than one perspective of trend at a time. This type of chart also provides a much better feel for the market's current level of price volatility. As a final example, I have drawn boxes around each weekly group of 4 so that we can see the daily inside the weekly and the weekly inside the monthly bars. This type of charting gives you a much different feel for the markets and their underlying trends.

It is also very interesting to note how often bars of similar size and/or harmonic proportion develop in the markets. Typically, as the markets lose directional momentum, the size of the bars begin to compress and decay.

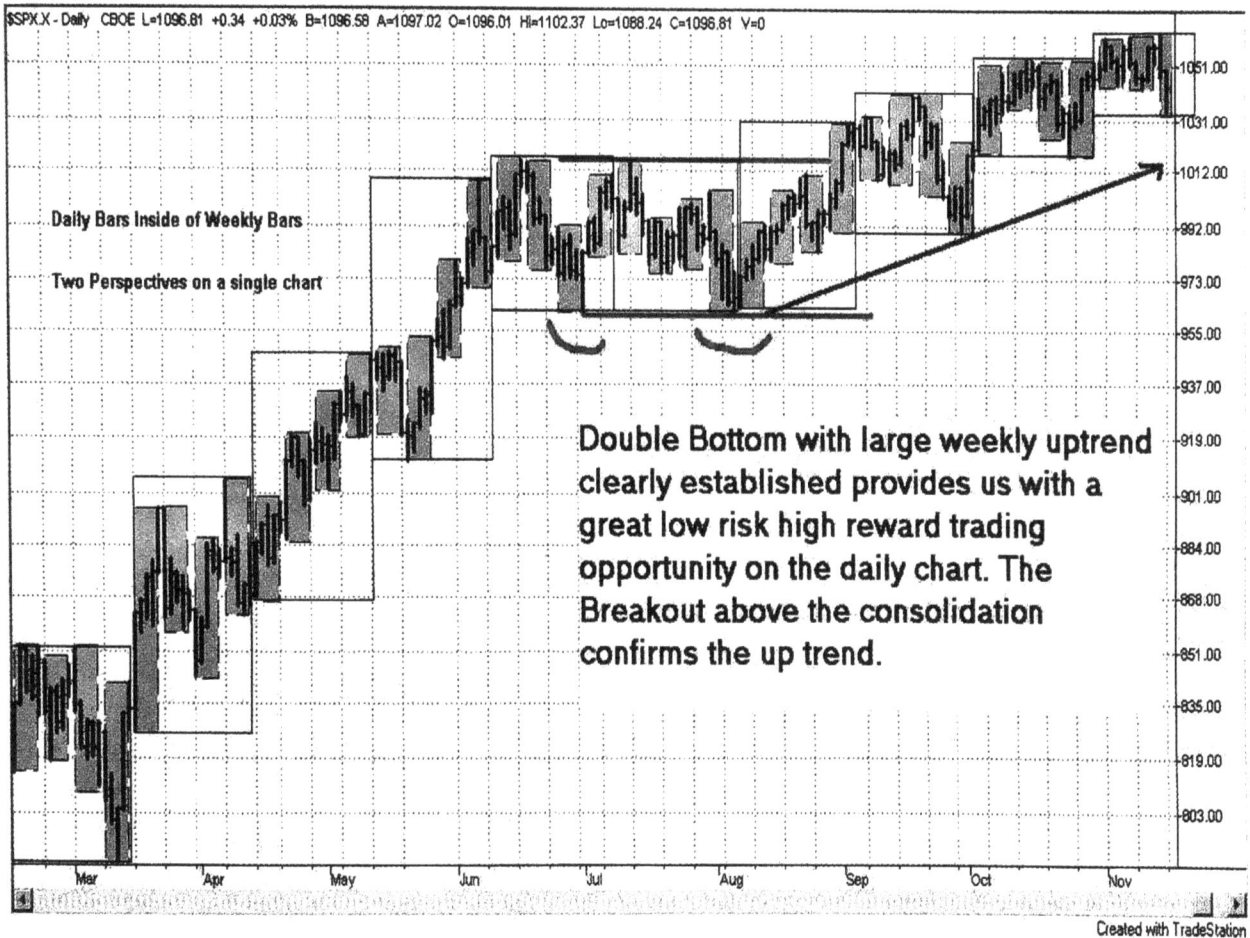

33

Therefore, there are two types of market trends that you can follow much more easily with this type of charting. (1) The directional trend of the price movement and (2) Volatility trends, which are based on the vertical size of the bars that result from grouping.

The next chart illustrates how to properly draw trend lines based on bar groups.

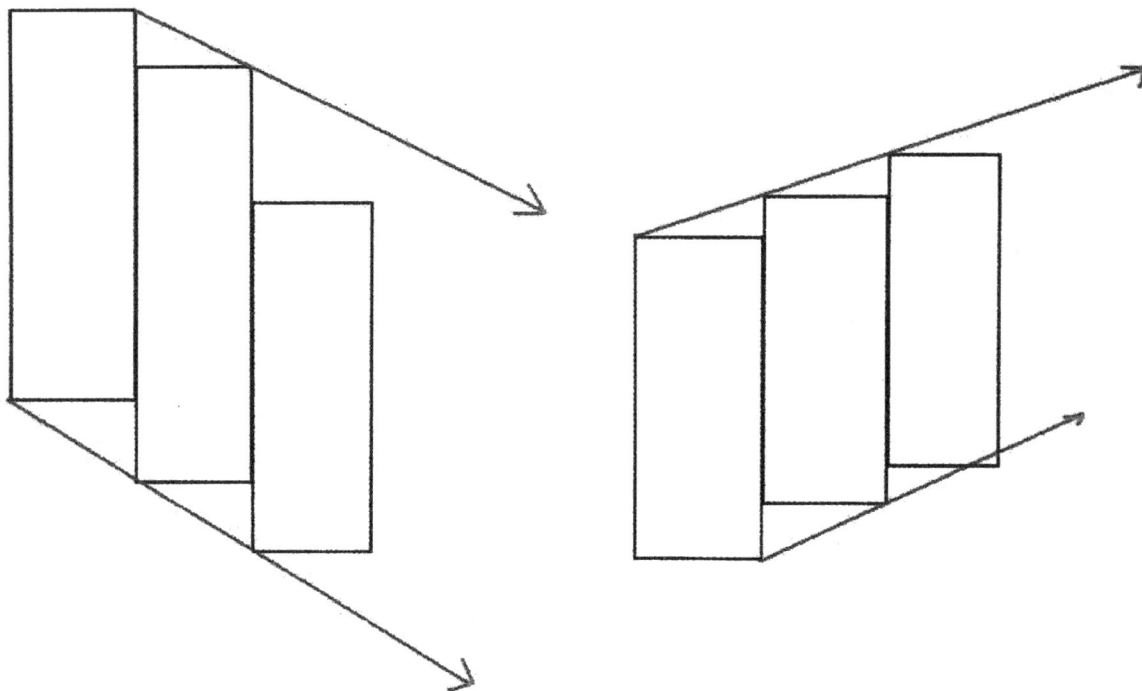

How to draw trend lines based on bar groups

If you would just visualize each bar group as a simple rectangle, the following description along with the above illustration should make the procedure perfectly clear. For **DOWN TRENDS**, start at the upper right corner of the rectangle and connect the trend to other upper right hand corners of the future bar groups. Also, project the lower channel of the trend by starting in the lower left hand corner of the rectangle and connecting the trend to other lower left corners of the future bar groups. For **UP TRENDS**, start at the lower right hand corner of the rectangle and connect the trend to other lower right hand corners of the future bar groups. In addition, project the upper channel by starting at the upper left hand corner of the rectangle and connecting the trend to other upper left corners of the future bar groups. The trends can only be projected from bar groups of the same time scale. For example, only connect weekly groups to other weekly groups or monthly groups to other monthly. Do not mix up the scales when drawing grouped trend lines. Often times, you can pin point near exact tops and bottoms where the particular market will meet with very strong support or resistance through the use of these trend lines. You simply determine how "tall" the next bar would have to be in order to hit the upper and lower trend line and that will give you a specific price that will act as support or resistance for the next bar group. If you based your trend lines on weekly groups, then the numbers will be weekly support and resistance. If you base your trend lines on monthly groups, then the prices will be monthly support and resistance figures. The next chart illustrates the approach.

Measure the vertical distance to the upper and lower trend lines and then project out horizontally from this exact point to determine prices that will act as strong support or resistance for the next bar group.

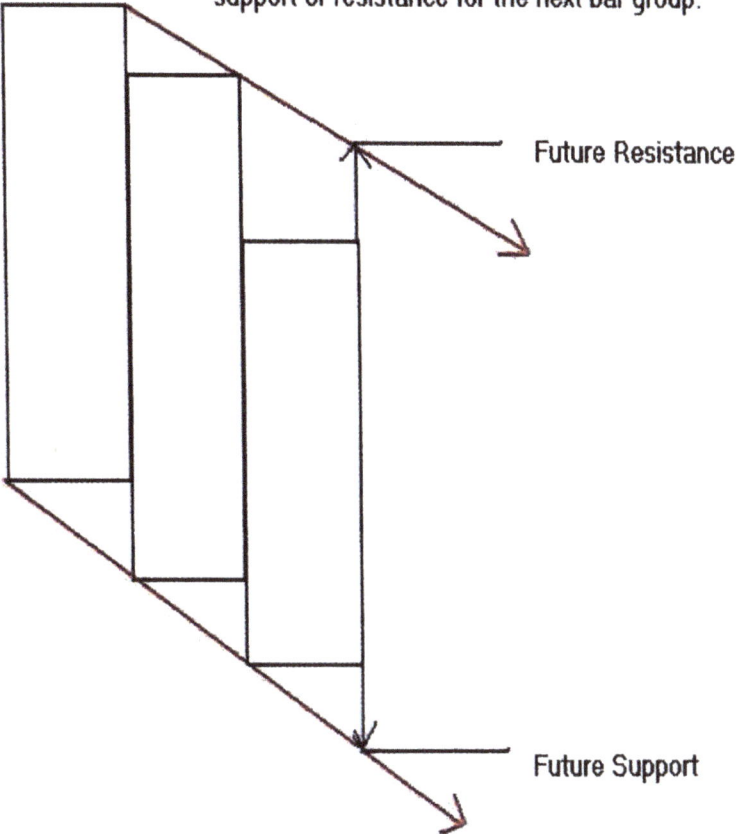

Future Resistance

Future Support

SWING TRADING

A QUICK WORD ABOUT W.D. GANN

W.D. Gann is reputed to have made over $50,000,000.00 over his 52-year career of trading in the stock and commodity markets. It has been reported that Gann started trading with $300 and made over $25,000 in profits his very first year of active trading. He then took $973 and made over $30,000 in the cotton market in 60-days time. Over a period of years, on each $1,000 he started with, he made profits of $26,500. A close friend of Gann said, *"He has made a great deal of money in the markets. I once saw him take $130, and in less than a month, run it up to over $12,000. He has taken half a million dollars out of the markets in the past few years. He can compound money faster than any man I have ever met "* It is important to understand that Gann was a swing trader and that he considered the swing trading approach to be the most profitable method of trading the markets. On page 132 of the *New Stock Trend Detector,* Gann writes, *"There is always more money to be made trading for fluctuations than holding for the long pull. A study of the swings in active stocks will convince a man that if he can get 25 to 50% out of each swing of 8 points or more, he can make far greater profits than in any other way of trading."* Of course, in typical Gann fashion, he also made a contradictory statement in many of his other writings that basically said that trading for the long pull and pyramiding the profits is the most profitable method of trading. However, a thorough review of his trading advice and "mechanical methods" will show that a long pull is just a larger swing trade when you understand the basics of his trading approach. This is why Gann consistently said to maintain a daily, weekly, monthly, and yearly chart of the market you were trading in. The "long pull" trades are generated using the same basic trading logic on a much longer time frame chart. That said; let's review Gann's exact words regarding trading and his swing trading method.

GANN'S MECHANICAL METHOD AND TREND INDICATOR

I am going to retype much of this method verbatim to maintain the trading principles involved, but you should understand that Gann had a couple of versions of this swing trading approach for various markets. Gann had a version for stocks, a version for grains, a version for cotton and coffee, etc. Therefore, I am focusing most of the attention on what they all have in common and what the main trading principles involved in the system are.

CAPITAL REQUIRED

The first and most important thing to know when you start to trade is the amount of capital required that will enable you to continue to trade 1, 2, 3, 5 or 20 years and never lose your capital. If you always have capital to start trading again, you can recoup losses and make money, but if you risk all of your capital on one or two trades and lose it, then the chances are against you. "Safety first" is the rule to apply. There is one safe, sure rule, and the man who will follow it and never deviate from it will always keep his money and come out ahead at the end of every year. This rule is; divide your capital into 10 equal parts and never risk more than this on any one trade. If you take a loss, then divide your remaining capital by 10 again and never risk more than this amount. As long as you have capital to operate with, you can always find new opportunities to make profits. Under this strict rule, you would have to lose more than 10 consecutive times before your capital would be wiped out. In my experience this has never happened and never will happen if you trade according to the rules and buy and sell on the trend line bottoms and tops, double or triple bottoms and tops and the 50% reaction or half way points.

KIND OF CHARTS YOU SHOULD KEEP

You should always keep up and maintain a daily open, high, low, close chart with your trend line indicator plotted on it. You should also keep a 3-day chart, a weekly chart, and a monthly chart in the same way. The monthly chart should go back 15 to 20 years so you will know where the important tops and bottoms are and where old resistance levels have been made before. This is often very helpful.

CHARTING THE TREND LINE INDICATOR

Keep up all of your charts and mark the trend line indicator each day. The daily trend indicator or trend line is obtained by following the daily price moves or bars. As long as the market is advancing, making higher bottoms and higher tops, the indicator or trend line is moved up each day to the highest price reached and continues to move up as long as the market makes both higher tops and higher bottoms. The first day the market makes both a lower bottom and a lower top, the trend line is moved down to the lowest price and continues to move down until the market can make both a higher top and a higher bottom. This trend line indicator simply follows the swings of a market.

I know this may sound a little confusing to some, so I will try to explain it as a simple math formula.

(1) If today's high is greater than yesterdays high and today's low is greater than yesterday's low, move the trend line up to the highest high.
(2) If today's low is less than yesterday's low and today's high is less than yesterday's high, move the trend line down to the lowest low.
I programmed this simple logic into TradeStation as a paintbar study that would color the bars green when the trend line should be moving up and red when the trend line should be moving down. You should plot a trend line indicator on the weekly, monthly and yearly charts as well, if you choose to follow Gann's method.

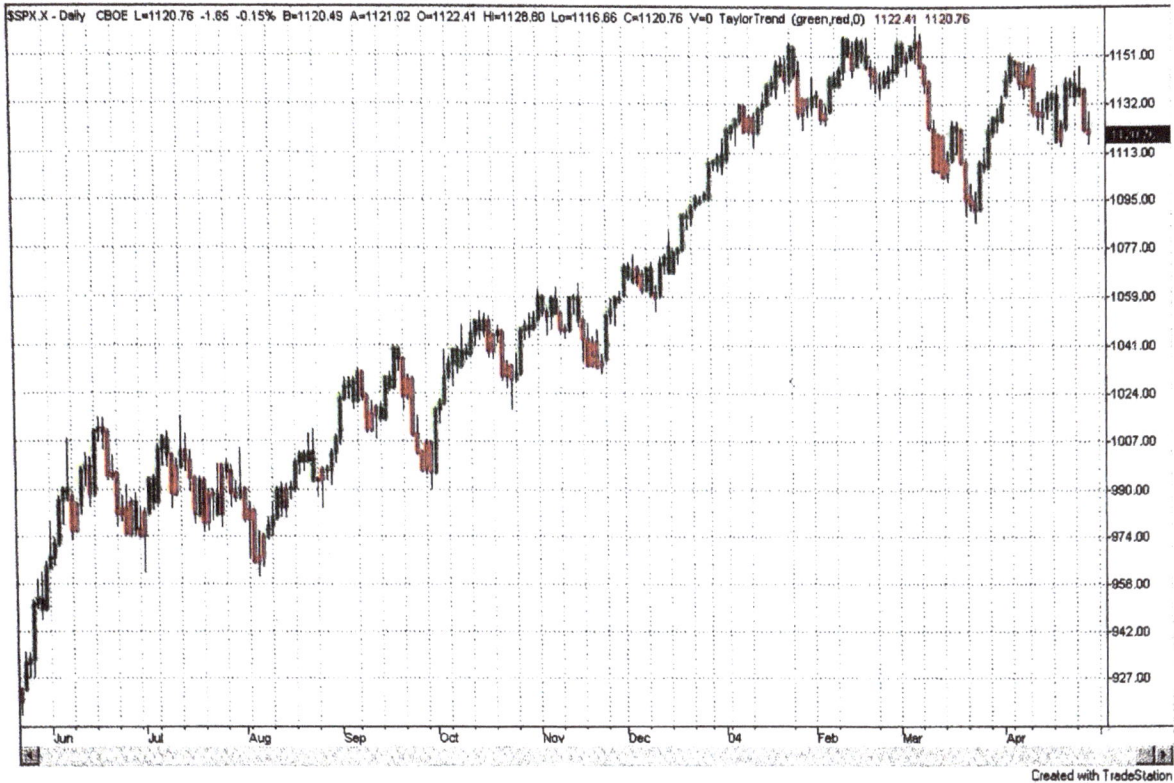

Next, I will manually draw in the trend line indicator, so you can see a clear example of what Gann was describing.

Now, it should be pretty clear that this indicator simply follows the swings in a market.

HOW TO USE STOP LOSS ORDERS

With this method you must always use a stop loss order 1, 2, or 3 points below the bottom or above the tops made by the trend line indicator. Remember that stops are placed just above the last swing top or just below the last swing bottom made on the trend line indicator and NOT placed above or below the high and low for the day. The amount of risk, if you were to get stopped out, must never be greater than 10% of your remaining trading capital.

TRADING INSTRUCTIONS FOR THE BUYING & SELLING POINTS

Double Tops

Triple Top

Double Bottoms

Triple Bottom

RULE #1

(1) The simplest and easiest rule to use is to buy long on a weekly or monthly chart's trend line double bottom or triple bottom and place your stop loss order just a few points under the last trend line bottom. Then on the daily chart continue to follow the market up with a stop loss order 1 to 3-points under the last daily trend line bottom and never use any other indication to sell out your position until your stop is caught and the prior trend line bottom is violated.

(2) The second easiest rule is to sell short when the weekly or monthly trend line indicator forms a double or triple top and place your stop loss order a few points above the prior trend line top. Then on the daily chart continue to follow the market down with a stop loss order 1 to 3-points above the last daily trend line top and never use any other indication to close out your short sale position until your stop is caught and the prior trend line top is violated.

RULE #2 DOUBLE OR TRIPLE TOPS & BOTTOMS

On the daily trend line indicator, buy against double or triple bottoms and protect with a stop loss order 1, 2, or 3 points away according to recent price activity. When a market makes the same price level a few days apart, it makes what we call a double bottom on the trend line indicator. A triple bottom is when a market makes a bottom around the same level of prices the third time. The second or third bottom can be slightly higher or slightly lower than the previous bottom, but always remember this rule:

When you buy at the time a market reaches the third bottom, you should never risk more than 1-point, for when the third bottom is broken, especially if this bottom is around the same price level, and your stop is caught, it generally indicates that the main trend has changed and you should prepare to go short.

This next rule is just the reverse of rule #2. Sell short against double or triple tops with a stop loss order 1, 2, or 3 points above the prior daily trend line top. On the third top, never risk more than 1-point, for when the third top is broken, it generally indicates higher prices and a change in trend.

The safest buying and selling points on the daily trend line indicator are when a market makes a triple top or triple bottom near the same price level and protecting the position with a stop loss order of 1 point above or below the prior swing.

RULE #3 FOURTH TIME AT THE SAME LEVEL

It is very important to watch a market when it reaches the same price level the fourth time as it nearly always goes through it on the 4th attempt. Therefore, it is safe to enter or reverse positions when a market goes through a triple top or triple bottom. Then continue to follow the market up or down on the daily chart with the daily trend line indicator moving your stops according to Rule #1.

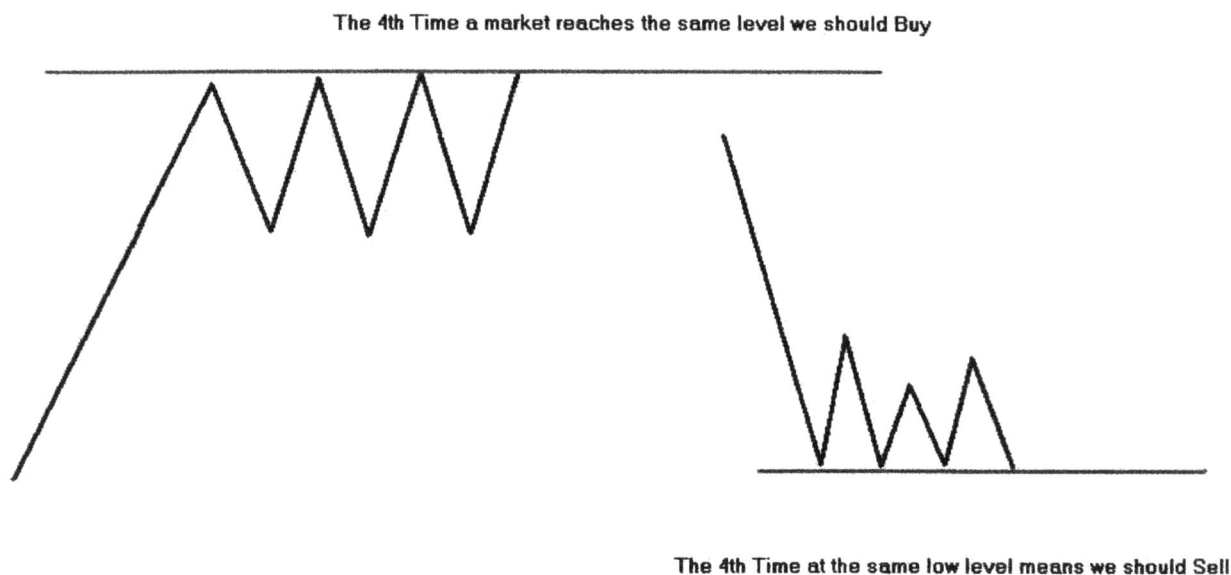

The 4th Time a market reaches the same level we should Buy

The 4th Time at the same low level means we should Sell

RULE #4 THE AVERAGE OR HALFWAY POINT

Always remember that the 50% reaction or halfway point of the range of fluctuation or of the extreme highest price of a market or of any particular movement is the most important point for support on the down side or for meeting resistance on the way backup. This is the balancing point because it divides the range of fluctuation into two equal parts or divides the highest selling price into two equal parts. For example, assume the market has a low at $40 and a high at $132. The two most important prices in order of strength would be $66, or 50% of the highest selling price, and $86, which is the halfway point between $40 and $132.

When a market advances or declines to the halfway point, you should always sell or buy with a stop loss order 1, 2, or 3 points away. The wider the price range and the longer the time period, the more important this halfway point is when it is reached. For example, if the market mentioned above declined from $132 to $86, we would buy long with a stop loss a few points away and then follow the market with the trend line indicator. If the market declined to $80 and then attempted to advance, we would look to sell short at $86 with a stop loss a few points above. We would react the same way at $66, buying long the first time this price was reached in the decline from the high of $132. If the market goes lower and attempts to advance back up to this level, we would sell short against it and follow the market with our stops using the trend line indicator as we did in Rule #1.

The greatest indication of strength is when a market holds one or more points above the halfway point, which shows that buying and support orders were placed above this important price level.

A sign of weakness is when a market advances and fails to reach or penetrate the halfway point by 1, 2, or 3 points and then declines and breaks the trend line indicator or other price resistance levels.

Always remember to calculate the 50% reaction price of the highest price and the 50% reaction of the low to high price for any stock, commodity, or market. This is the balancing point because it divides the price and it divides the range into two equal parts. The wider the range is between high and low and the longer the amount of time that has elapsed, the more important these points are when reached.

You can make a fortune by following this one rule alone. A careful study and review of past movements in any stock, commodity, or average will prove to you beyond doubt that this rule works and that you can make profits following it. This is a rule that can make you rich.

LOST MOTION

As there is lost motion in every kind of machinery, so there is lost motion in the markets due to momentum, which drives a market slightly above or below support or resistance levels. The average lost motion is 1 to 3 points.

When a market is very active and advances or declines fast on heavy sales, it will often go from 1 to 3 points above or below a halfway point or other strong resistance levels but will not go over 3 full points beyond it.

This is the same rule that applies to a gravity center in anything. If we could bore a hole through the earth and then drop a ball, momentum would carry it beyond the gravity center, but when it slowed down, it would finally settle on the exact center. This is the way markets act around these important price centers.

This mechanical method, based on fixed rules, will overcome the human element, eliminating hope and fear as well as guesswork. Learn to stick to the rules and trade in active markets. Remember, practice makes perfect. The more work and study you do the greater success you will have. Apply all of the rules, keep your charts up to date at all times, and do no guesswork, and your success is sure. In order to make a success in trading the markets, the trader must have definite rules and follow them.

The rules given below are based upon my personal experience and anyone who follows them will make a success.

* Divide your capital into 10 equal parts and never risk more than 1/10th of your capital on any one trade.
* Always use stop loss orders to protect your trade and place it immediately after entering a position.
* Never overtrade by taking large positions. This would violate your capital rule. Remember "safety first."
* Never let a profit run into a loss. When the market moves in your favor and you have a profit that is double the amount of risk you were willing to take, move your stop loss order so that you will have no loss of capital if hit.
* When in doubt, stay out or get out.
* Trade in active, liquid markets.
* Don't close your trades without a good reason. Follow up the position with stop loss orders to protect your accumulating profits according to the rules.
* Accumulate a surplus of capital. This rule is very important. After you have made a series of successful trades, put some money into a surplus account to be used only in an emergency or in times of panic. *I personally suggest that you withdraw 50% of your profits each time you close a successful trade, DTF*
* Never average a losing position. This is one of the worst mistakes a trader can make.
* Never get out of the market just because you have lost patience or get into a market because you are anxious from waiting.
* Avoid taking small profits and large losses.
* Never cancel a stop loss order after you have placed it when entering a trade.
* Avoid trading too frequently, getting in and out too often.
* Be just as willing to sell short as you are to buy.
* Never change your position without a good reason based on set rules.
* Avoid increasing your trading activity after a long period of success or a series of profitable trades. Remember to accumulate a surplus and don't be tempted to increase your trading unit too quickly. Success can go to your head and has ruined many otherwise good traders.

When you decide to make any trade, be sure that you are not violating any of these rules, which are vital and important to your success. If you close a trade with a loss, go over these rules and see if it was from violating any one of them. If so, then do not make the same mistake the second time. Experience and investigation will convince you of the value of these rules, and observation and study will lead you to a correct and practical theory for success. **—W.D. Gann.**

As mentioned before, Gann produced many versions of this "mechanical method." What I presented above is basically the common denominators that remained consistent in all of the various versions. Gann had some additional trading rules that I removed entirely because they were inconsistent with his money management rules. One of the rules I removed was a stop and reverse order. This rule frequently appeared in his trailing stop approach in Rule #1 and also in his Rule #2 for triple tops & bottoms. What Gann said to do in #1 originally was if your trailing stop was hit, you would immediately reverse direction and go with it. For example, if you were long 1 unit and placed your stop just below the prior swing low, you would have this stop order prepared as "sell 2-units" so that if it were hit, you would be net short 1 trading unit in the market. This particular rule or approach would have you in the market 100% of the time after your first entry. I have tested this rule and proven to myself that it simply does not work in today's markets. Gann also advised to use a stop and reverse order above and below triple tops and bottoms. This stop and reverse rule is not quite as bad as the first, but overall I still advise against using it. It is much safer and easier to trade these swing formations as they are and just take the gain or loss, win or lose. The markets will always provide another opportunity down the road. Gann also had rules for adding contracts as profits accumulated in the open position and building a pyramid. I also advise against this practice for the average trader, which is the reason it was removed from the " method." This practice just continues to raise your cost basis in the trade, making your overall position closer and closer to current "at the market" prices. I personally prefer to have some breathing room in a trade to keep a level head.

EXPLAINING GANN'S 50% RULE

When you reach the halfway point between any price swing or 50% of the extreme high, you are in a location that is exactly balanced between the two magnetic price polarities. A few quick market examples will prove the usefulness of this phenomenon based on "natural laws." The chart below illustrates several 50% points in the S&P500 stock market index.

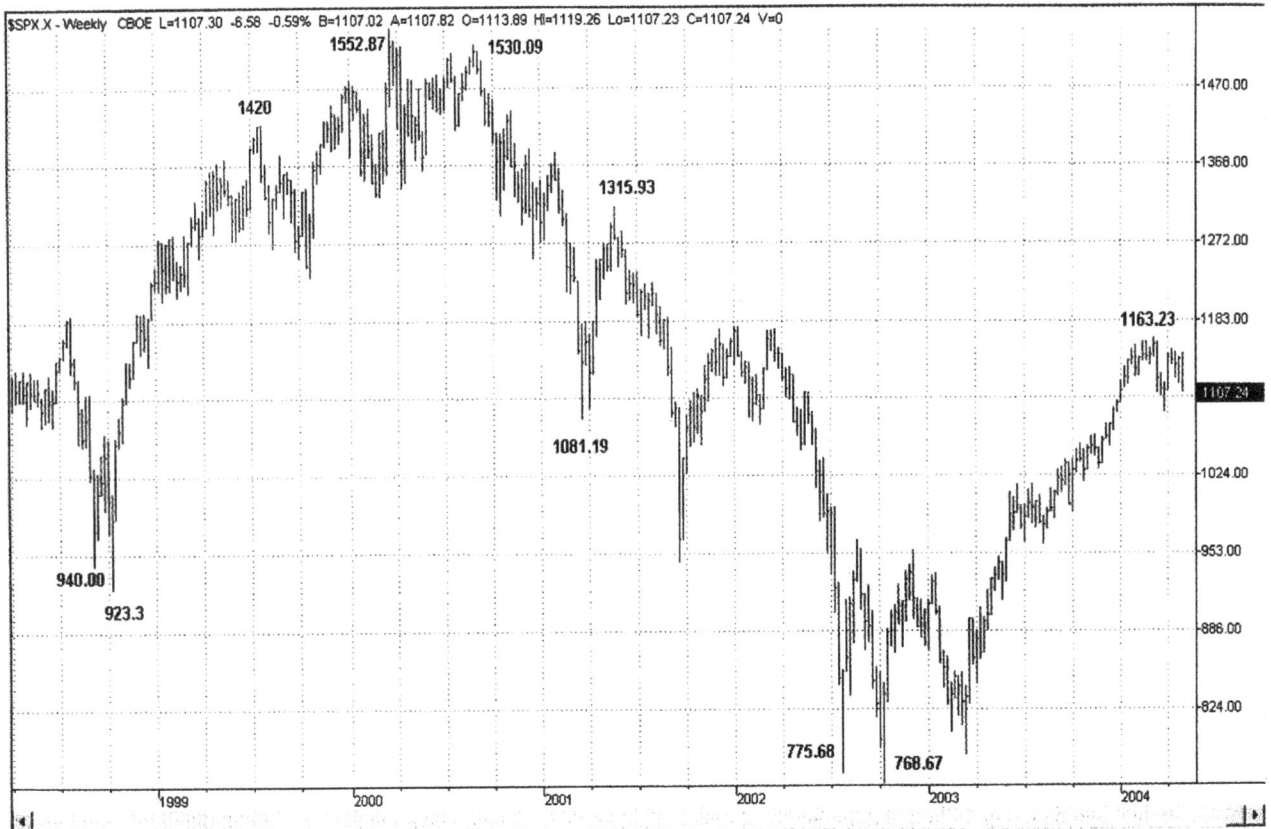

940.00 x 1.50 = 1410.00, actual top was 1420.
923.30 x 1.50 = 1384.95 (very close to the top that preceded 1420.)
The 50% point between the all time high 1552.87 and the low of 1081.19 =1317.03, the actual top was 1315.93.
50% of the all time high = 776.44. The first major low was 775.68.
50% of the secondary high of 1530.09 = 765.45, the lowest point reached ever was 768.67. (Try multiplying these two lows by 1.25 and see what you get!)
The halfway point between the highest and lowest prices is (1552.87 + 768.67)/2 = 1160.77. The top of this up swing reached 1163.23.
775.68 x 1.50 = 1163.52.

You should also look at 25% declines and or advances from price extremes as this would be the halfway point between the gravity center or 50% point. For example, 775.68 x 1.25 = 969.60, which is very close to the high reached on August 22, 2002. Subtracting 25% from the all time high (1552. 87) gives 1164.65.

There are other examples to be found, but I think this illustrates the basic tendency. Traders Press also has a book called ***The Trading Rule That Can Make You Rich*,*** by Edward Dobson, which deals entirely with Gann's 50% rule.
http://www.traderspress.org/detail.asp?product_id=5

GANN'S RED LIGHT GREEN LIGHT TREND INDICATOR

Many people often get confused as to how to draw this swing indicator. The rules are actually very simple and I will attempt to clear up any confusion you may be having with this trend indicator. As I discussed with the simple comparison formula on page 38:

(1) If today's high is greater than yesterday's high and today's low is greater than yesterday's low, you move the trend line up to the highest high.

(2) If today's low is less than yesterday's low and today's high is less than yesterday's high, you move the trend line down to the lowest low.

Once the trend line begins in a direction (up or down) it continues in that direction until you meet the above criteria. For example, if we are currently in a down trend we will continue to move the trend line down (or color the bars red) each day (or bar) until we have both a higher high and a higher low. Once we have both a higher high and a higher low, we switch the trend to up and this new up trend will continue until we have both a lower low and a lower high. That is all there is to it! Gann would draw his up swings in green and his down swings in red. Gann advised to buy long when the trend changed from down to up (red to green) and to sell short when the trend changed from up to down (green to red). This does not work well as a single stand-alone strategy, but when you combine it with the chart patterns and longer term trend lines available through bar grouping; you can generate some very impressive results. I will illustrate some actual examples after we cover one final swing-trading pattern.

Now that you understand how to draw a Gann style swing chart, I would strongly urge you to carry across the ending prices of each swing on your charts to get a better understanding of market action and support & resistance levels.

ABC's AND 123's

Before we look at our next market example, I want to present another swing trading pattern that is both useful and effective for generating reliable trade entries. This is basically the pattern that Gann advised to use as a stop and reverse, but we will only use it as an entry technique that only follows the direction of the longer-term trends. First, I will just present a simple illustration of how the pattern looks.

ABC's and 123's

Bearish pattern. Point B is a lower swing top. Sell Short when the market breaks below point A.

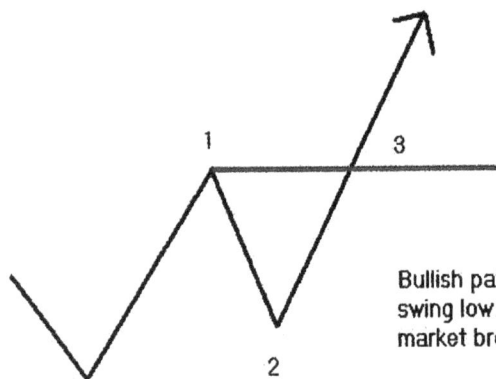

Bullish pattern. Point 2 is a higher swing low. Buy Long when the market breaks above point 1.

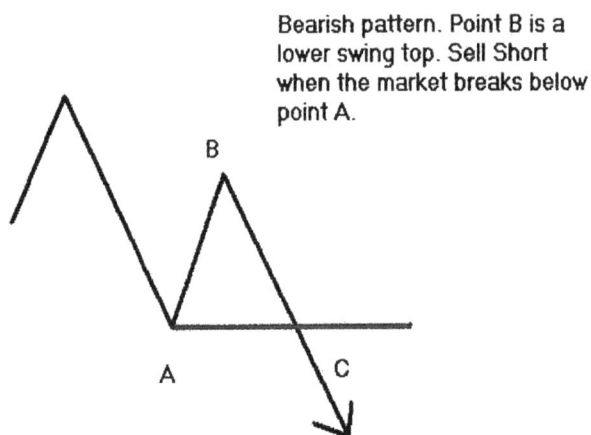

PUTTING IT ALL TOGETHER

The next series of charts will combine all the techniques discussed so far to aid with your understanding of the principles involved. I'm not one to waste time providing hundreds of examples of the same thing so please take time to study each chart carefully. I have found from private teaching that it is much more effective for the student to find his or her own examples after the basic concepts have been covered. This improves their visual recognition skills and pattern analysis, as well as their understanding of each technique. The daily chart of the S&P500 index below includes a weekly and monthly bar grouping as well as Gann's swing indicator. I have also used the paintbar study so that you can see when the trend changes color from green to red or vice versa. Also, understand that I drew everything on this chart manually other than the red and green painbars.

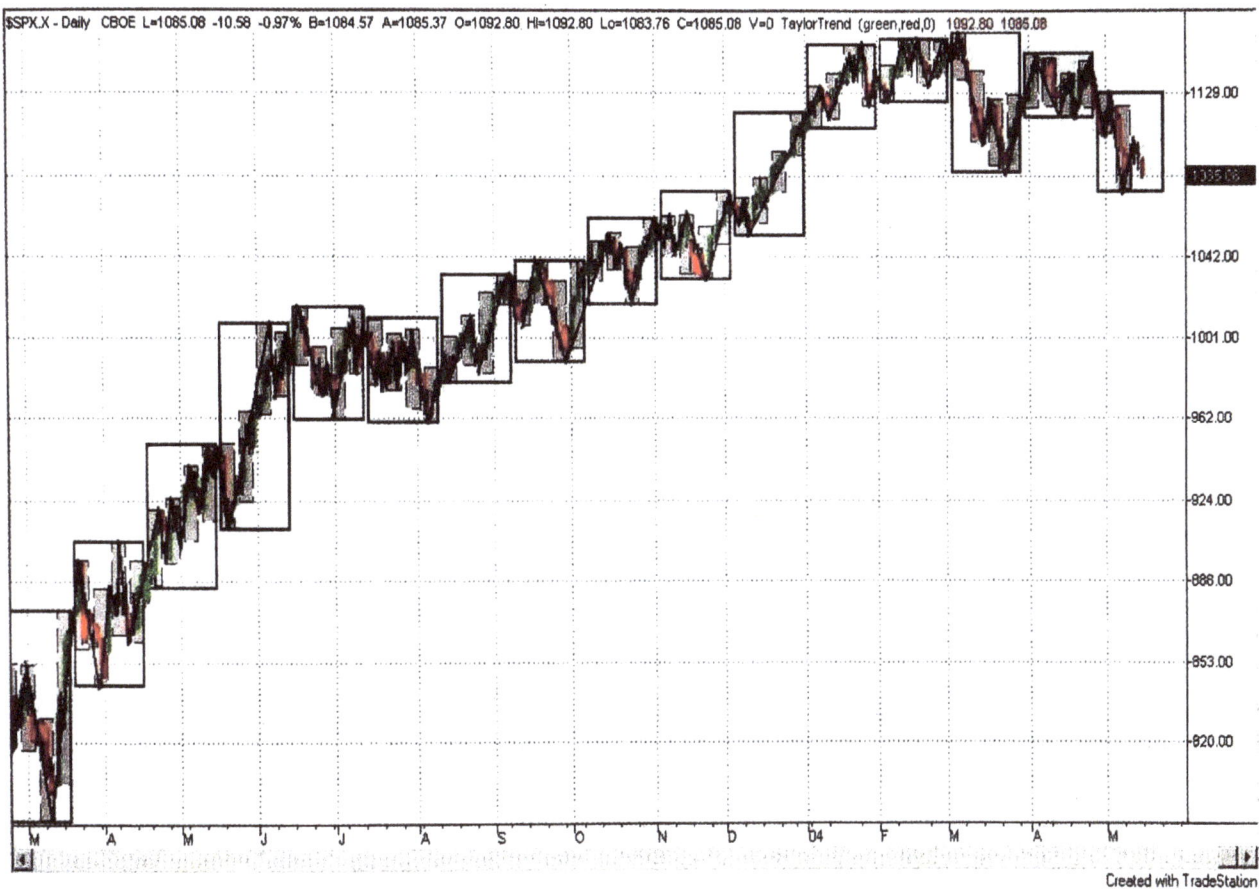

Now, let's look at the trades that were available to us from combining everything we have learned so far.

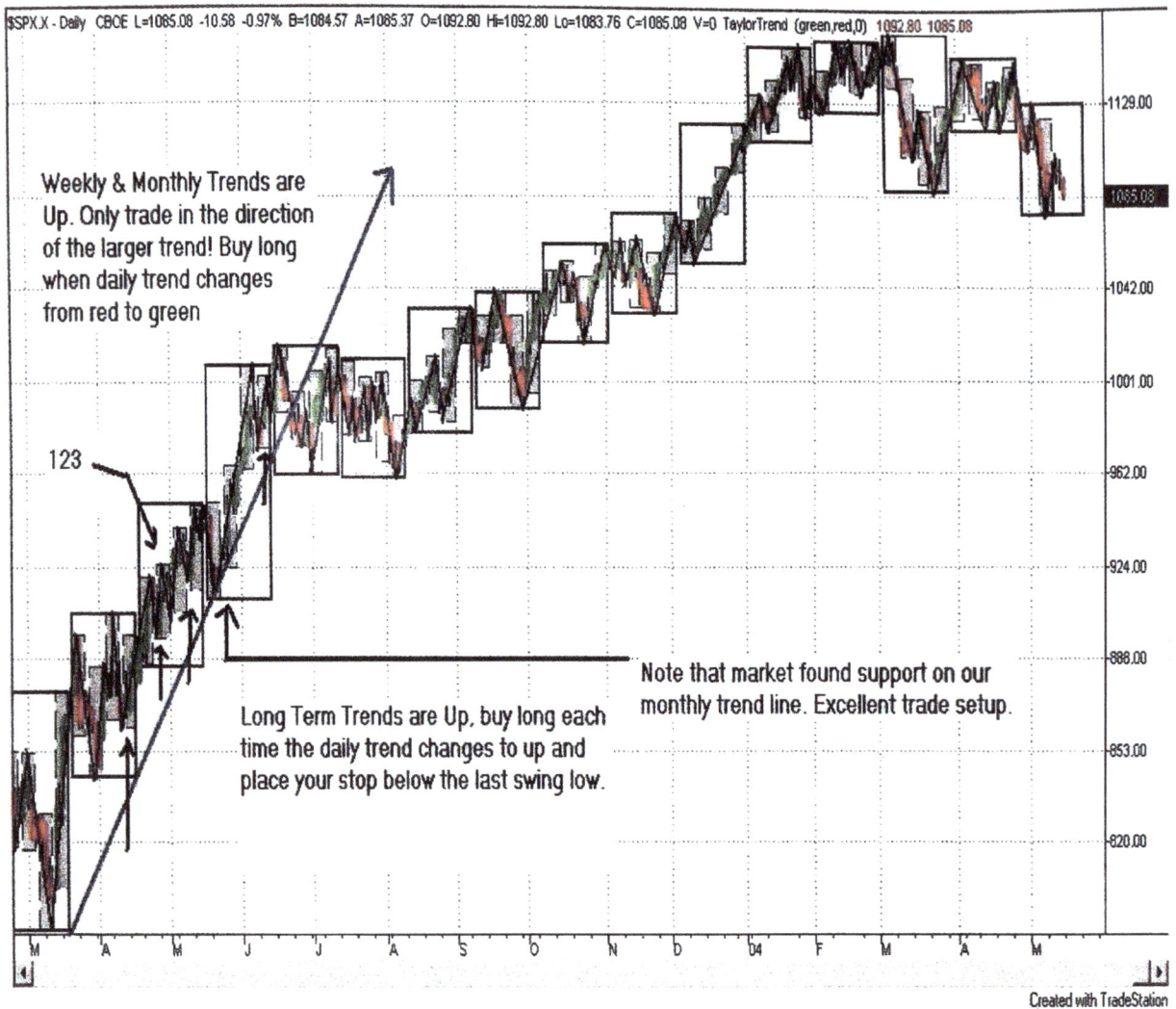

$SPX.X - Daily CBOE L=1085.08 -10.58 -0.97% B=1084.57 A=1085.37 O=1092.80 Hi=1092.80 Lo=1083.76 C=1085.08 V=0 TaylorTrend (green,red,0) 1092.80 1085.08

Weekly & Monthly Trends are Up. Only trade in the direction of the larger trend! Buy long when daily trend changes from red to green

123

Long Term Trends are Up, buy long each time the daily trend changes to up and place your stop below the last swing low.

Note that market found support on our monthly trend line. Excellent trade setup.

Created with TradeStation

Note that we are only trading in the direction of the larger weekly and monthly trends drawn from our bar grouping. These trends are very important to your trading success. As the saying goes, *"The Trend Is Your Friend."* Once the daily or weekly bars violate our trend, we immediately close out any open long positions. Now, since the trend has been broken and the future direction is in question, we no longer enter on the daily trend "red to green" change indications. The next chart will show the trades that were available based on our rules.

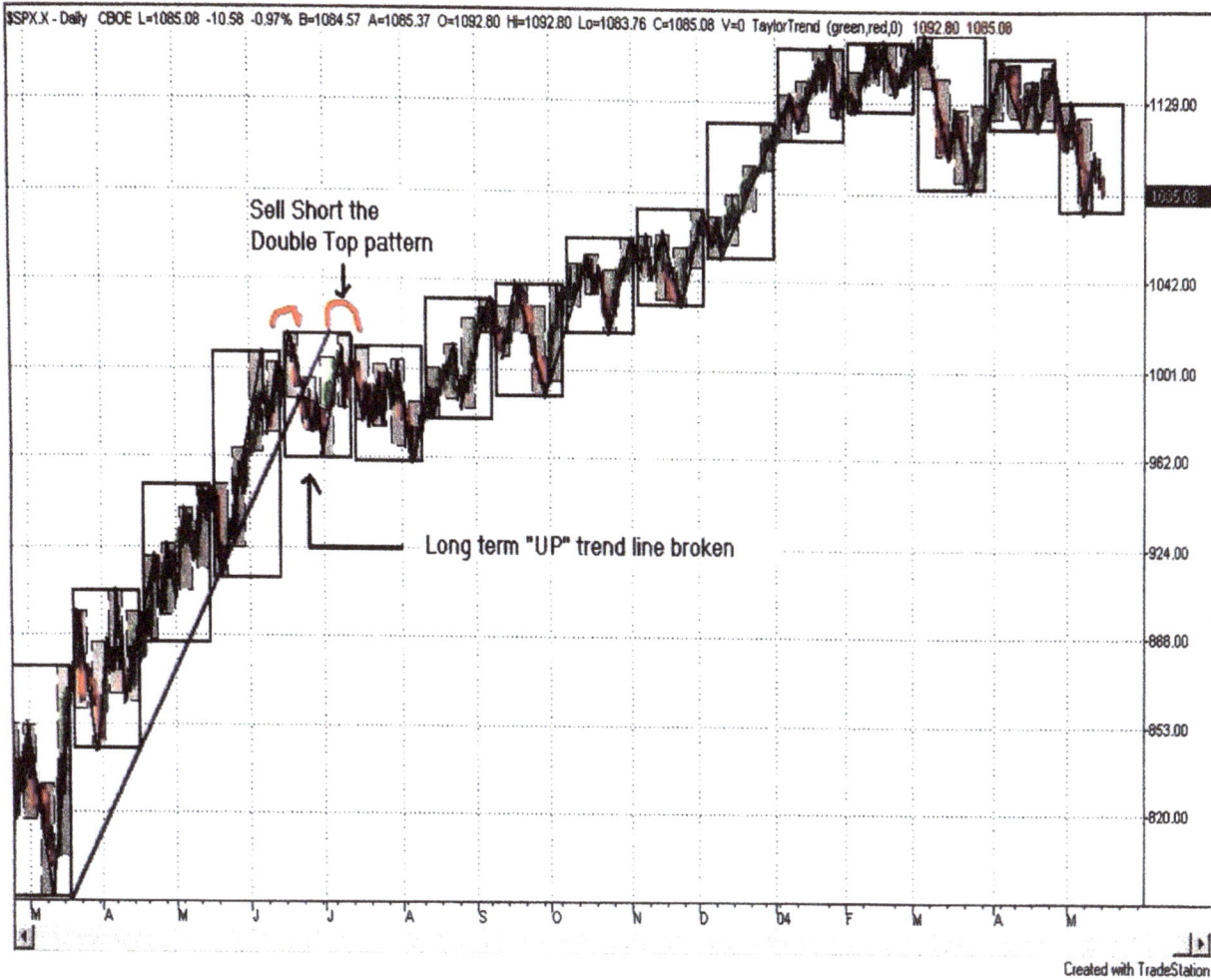

In this chart, we clearly see the double top formation that formed after breaking the longer-term trend providing a great low risk, high return opportunity. At this point, we still cannot utilize the green or red trend change signals from the daily indicator because we do not have any clear direction of trend. In other words, the longer-term trend is still questionable at this point in time.

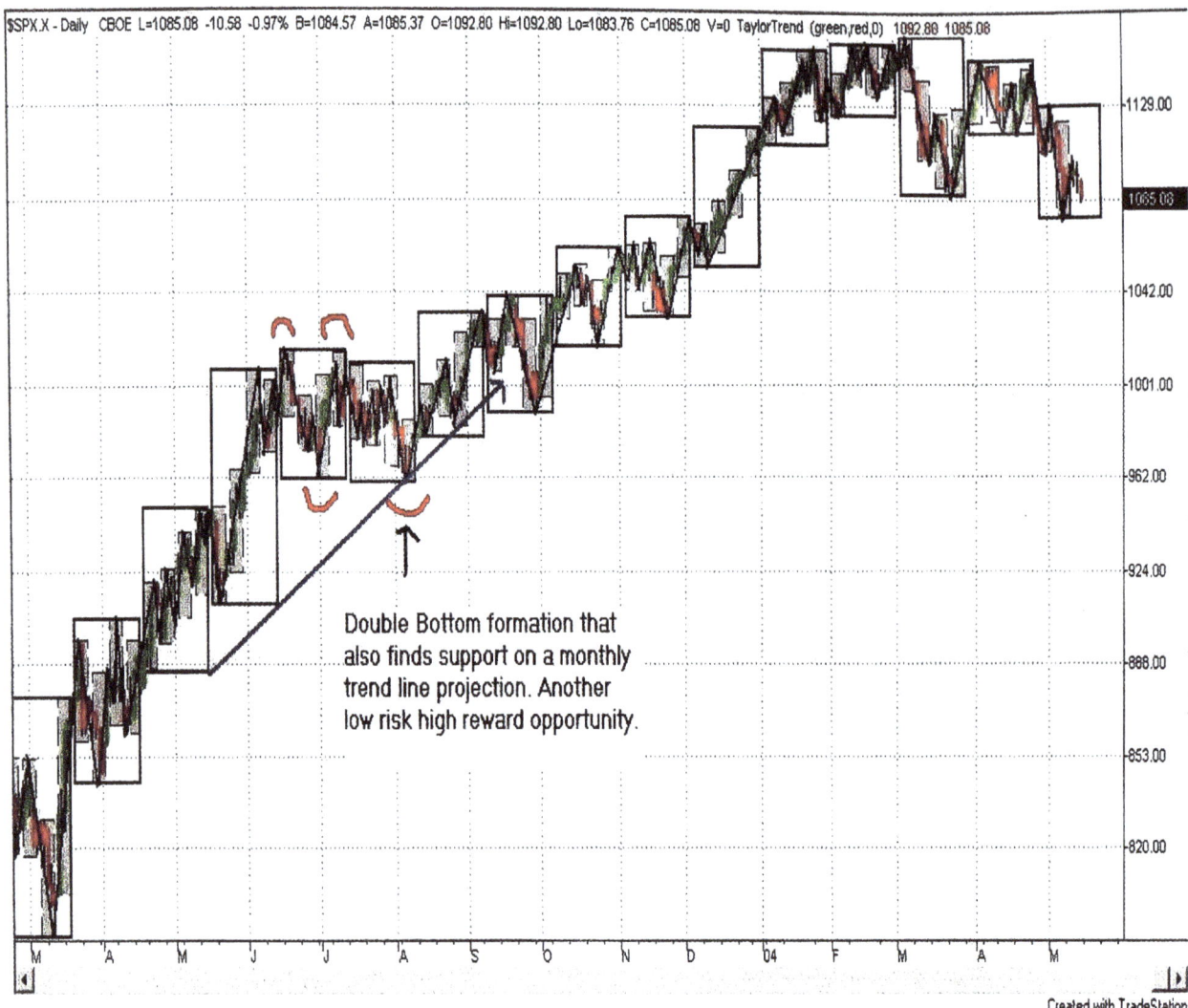

$SPX.X - Daily   CBOE   L=1085.08 -10.58 -0.97%   B=1084.57 A=1085.37 O=1092.80 Hi=1092.80 Lo=1083.76 C=1085.08 V=0   TaylorTrend (green,red,0)   1092.80 1085.08

Double Bottom formation that also finds support on a monthly trend line projection. Another low risk high reward opportunity.

Created with TradeStation

Now, after the double top formation, we are provided with a nice double bottom formation that finds support on a monthly trend line from our bar grouping. This is another great low risk, high reward trading opportunity.

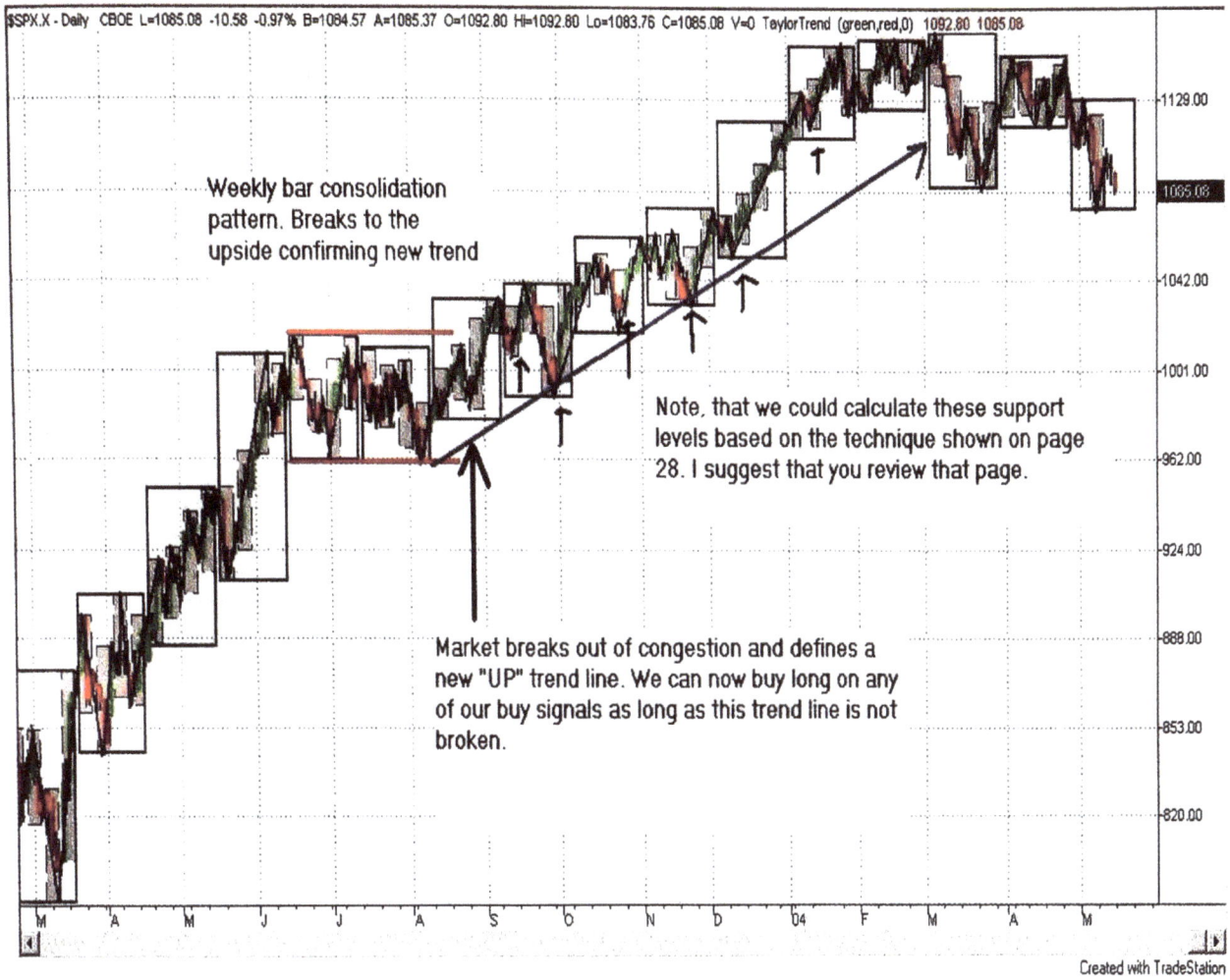

$SPX.X - Daily CBOE L=1085.08 -10.58 -0.97% B=1084.57 A=1085.37 O=1092.80 HI=1092.80 Lo=1083.76 C=1085.08 V=0 TaylorTrend (green,red,0) 1092.80 1085.08

Weekly bar consolidation pattern. Breaks to the upside confirming new trend

Note, that we could calculate these support levels based on the technique shown on page 28. I suggest that you review that page.

Market breaks out of congestion and defines a new "UP" trend line. We can now buy long on any of our buy signals as long as this trend line is not broken.

Created with TradeStation

Study this chart carefully. Also notice the triple top and 4th time at the same level pattern that developed on the daily swing chart as the market moved up.

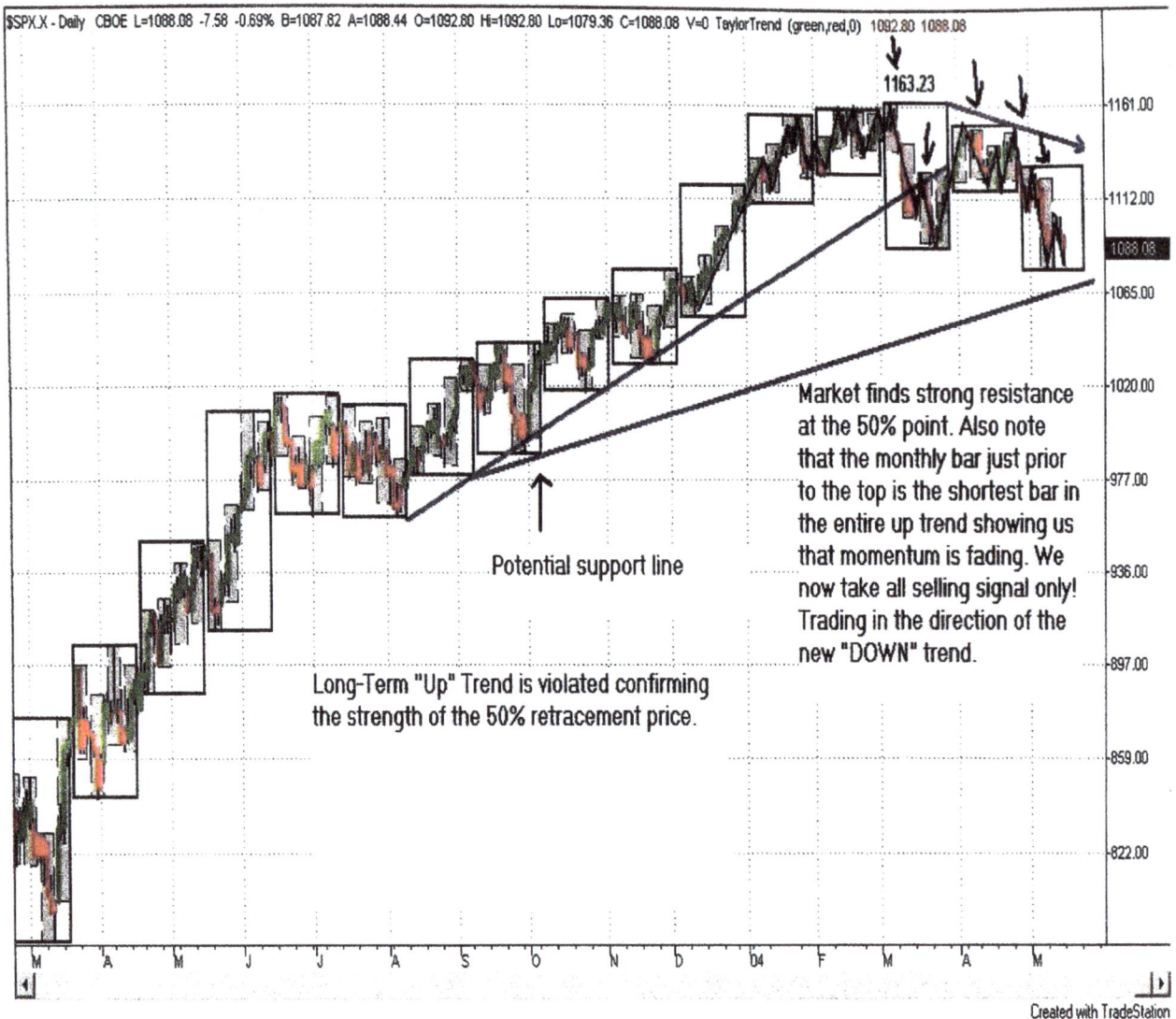

$SPX.X - Daily CBOE L=1088.08 -7.58 -0.69% B=1087.82 A=1088.44 O=1092.80 Hi=1092.80 Lo=1079.36 C=1088.08 V=0 TaylorTrend (green,red,0) 1092.80 1088.08

1163.23

Market finds strong resistance at the 50% point. Also note that the monthly bar just prior to the top is the shortest bar in the entire up trend showing us that momentum is fading. We now take all selling signal only! Trading in the direction of the new "DOWN" trend.

Potential support line

Long-Term "Up" Trend is violated confirming the strength of the 50% retracement price.

Created with TradeStation

On this chart, we get to see Gann's 50% rule in action. Please review page 46 for additional details. Here, the market finds strong resistance almost exactly on the 50% level, providing us with another relatively low risk, high reward opportunity. The market then trades down far enough to violate the monthly bar group up trend that had been in force since the congestion breakout. This trend violation after testing and failing at the 50% level is a solid confirmation that we are at a significant top in this market. Also, note the nice double top pattern indicated by our swing chart.

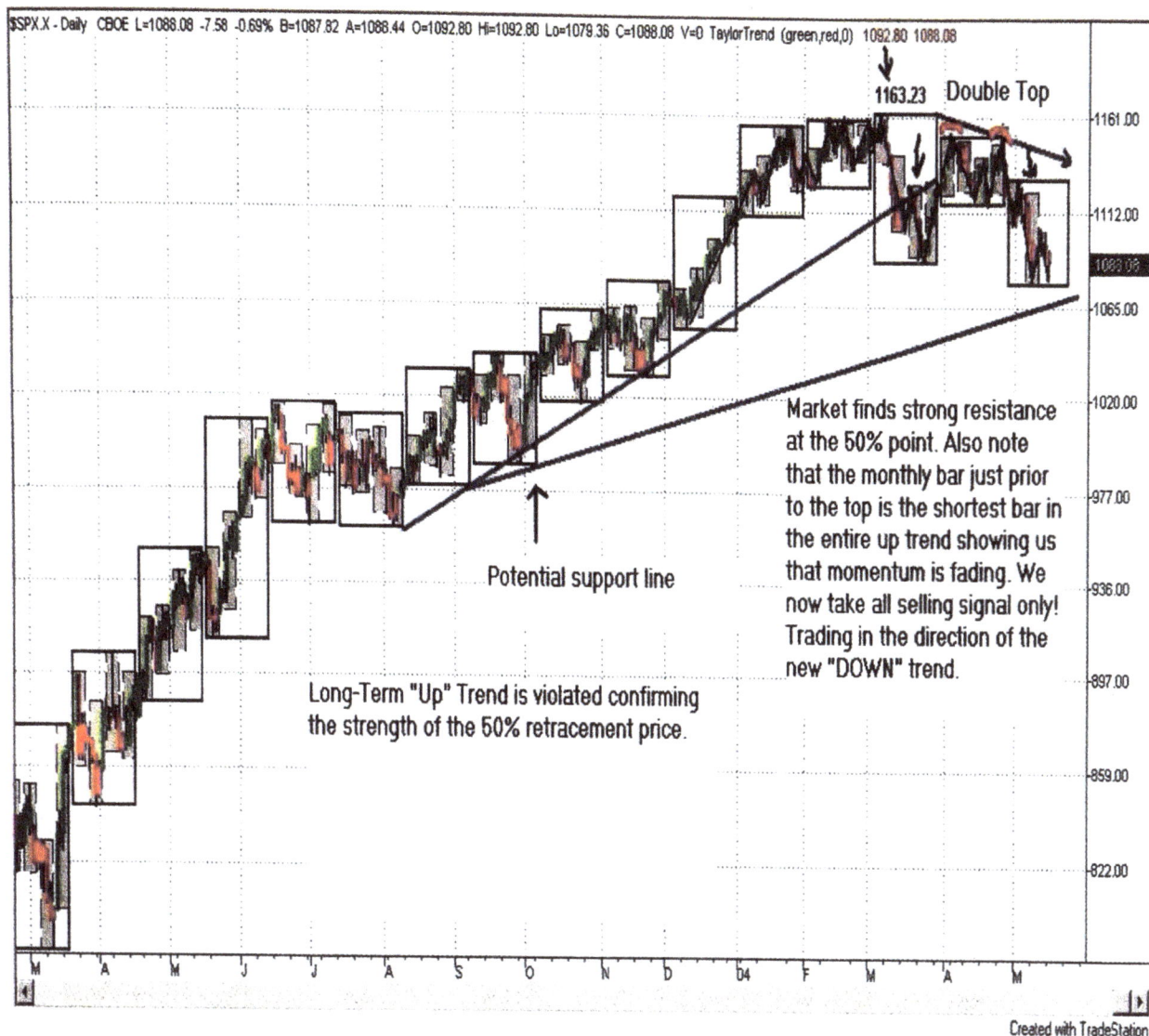

$SPX.X - Daily CBOE L=1088.08 -7.58 -0.69% B=1087.82 A=1088.44 O=1092.80 Hi=1092.80 Lo=1079.36 C=1088.08 V=0 TaylorTrend (green,red,0) 1092.80 1088.08

1163.23 Double Top

Market finds strong resistance at the 50% point. Also note that the monthly bar just prior to the top is the shortest bar in the entire up trend showing us that momentum is fading. We now take all selling signal only! Trading in the direction of the new "DOWN" trend.

Potential support line

Long-Term "Up" Trend is violated confirming the strength of the 50% retracement price.

Created with TradeStation

The double top pattern following the monthly "UP" trend violation and test & failure at the 50% price, provided another great short position entry into this newly forming down trend. Using the technique described on page 35, we can determine that the market would have to drop down to 1064.50 to hit the potential support line drawn on the chart above. The bar grouping trend line technique discussed on page 34, also highlights the significance of the 1064.50 price level. If this level is broken to the downside, the "bearish" trend could strengthen quite a bit. The next chart illustrates the geometrical importance of this price level.

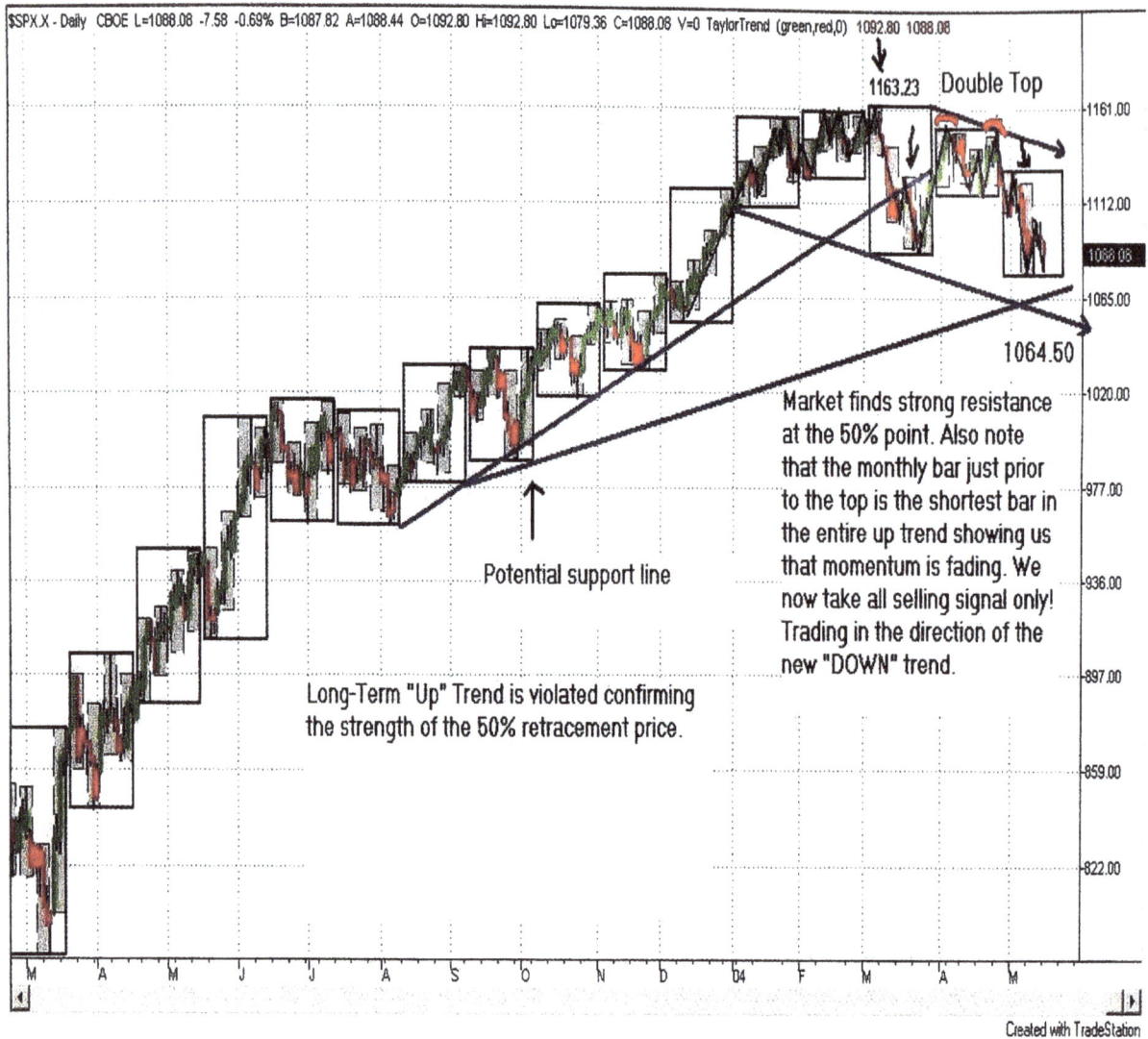

$SPX.X - Daily CBOE L=1088.08 -7.58 -0.69% B=1087.82 A=1088.44 O=1092.80 Hi=1092.80 Lo=1079.36 C=1088.08 V=0 TaylorTrend (green,red,0) 1092.80 1088.08

1163.23   Double Top

1064.50

Market finds strong resistance
at the 50% point. Also note
that the monthly bar just prior
to the top is the shortest bar in
the entire up trend showing us
that momentum is fading. We
now take all selling signal only!
Trading in the direction of the
new "DOWN" trend.

Potential support line

Long-Term "Up" Trend is violated confirming
the strength of the 50% retracement price.

Note that these two down trend lines are approximately parallel to one another. The lower line also intersects the potential support line at exactly the same price we calculated by extending the monthly bar down to the trend line. We would certainly exit all short positions at this price and look for support to enter the market here. This does not mean we enter long trades. We are still only interested in selling short as long as the down trend line remains in tact. Even if the trend line were to be broken, we would also look for a potential short sell trade entry at the 50% price level again, which would potentially form a double top pattern at this "**KEY**" price level.

We have covered a lot of ground in these few pages. You have been introduced to a unique multiple perspective approach allowing you to see trends more easily. In addition, we have combined this multi-perspective view with W.D. Gann's mechanical trading methods to produce a very detailed approach to swing trading. It is critical that you fully understand everything covered up to this point, money management being the most critical issue followed second by the trading rules. Remember, if you are diligent with your money management, you can be wrong more often than you are correct and still make money. In my opinion, this is the true reason Gann was successful in the markets.

Trading with the trend combined with sound money management can be extremely profitable. Before we move on to the final entry signal, we will quickly review our definition of trend.

Up Trends on bar charts have a series of higher highs and higher lows.

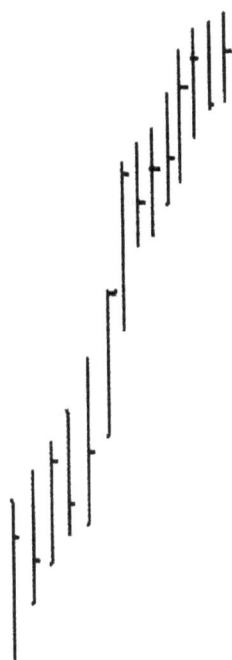

Up Trends can also be defined as a series of higher swing highs and higher swing lows on a swing chart.

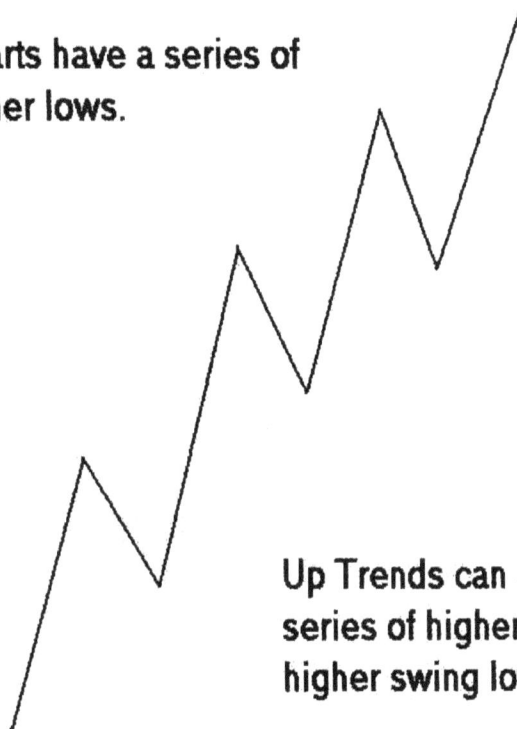

Down trends are a series of lower lows and lower highs on a bar chart. On a swing chart, we would see a series of lower swing highs and lower swing lows.

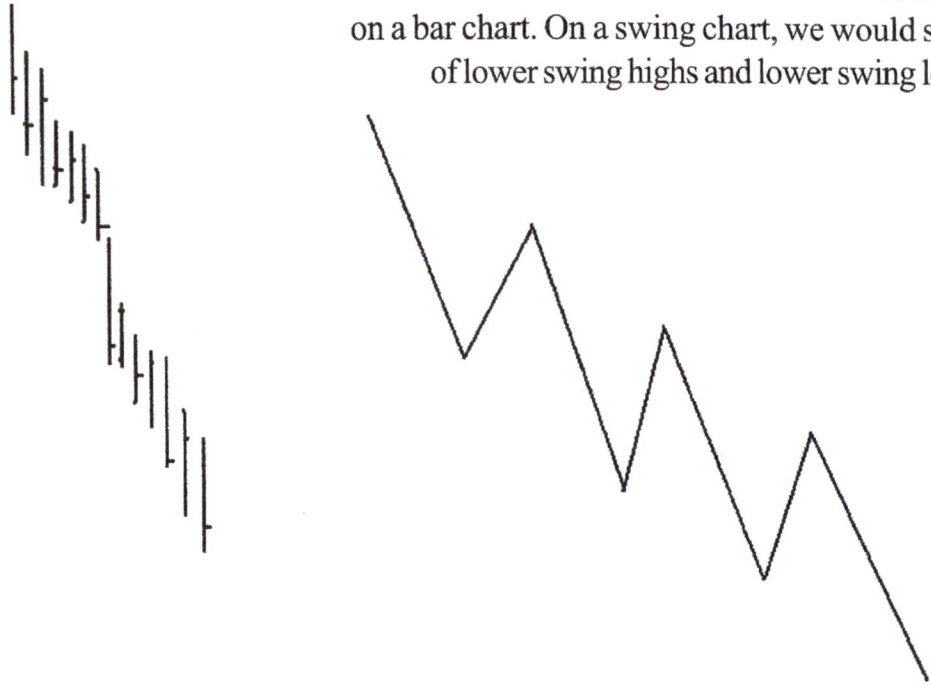

Now, quickly review these definitions and compare them to our multi-perspective view.

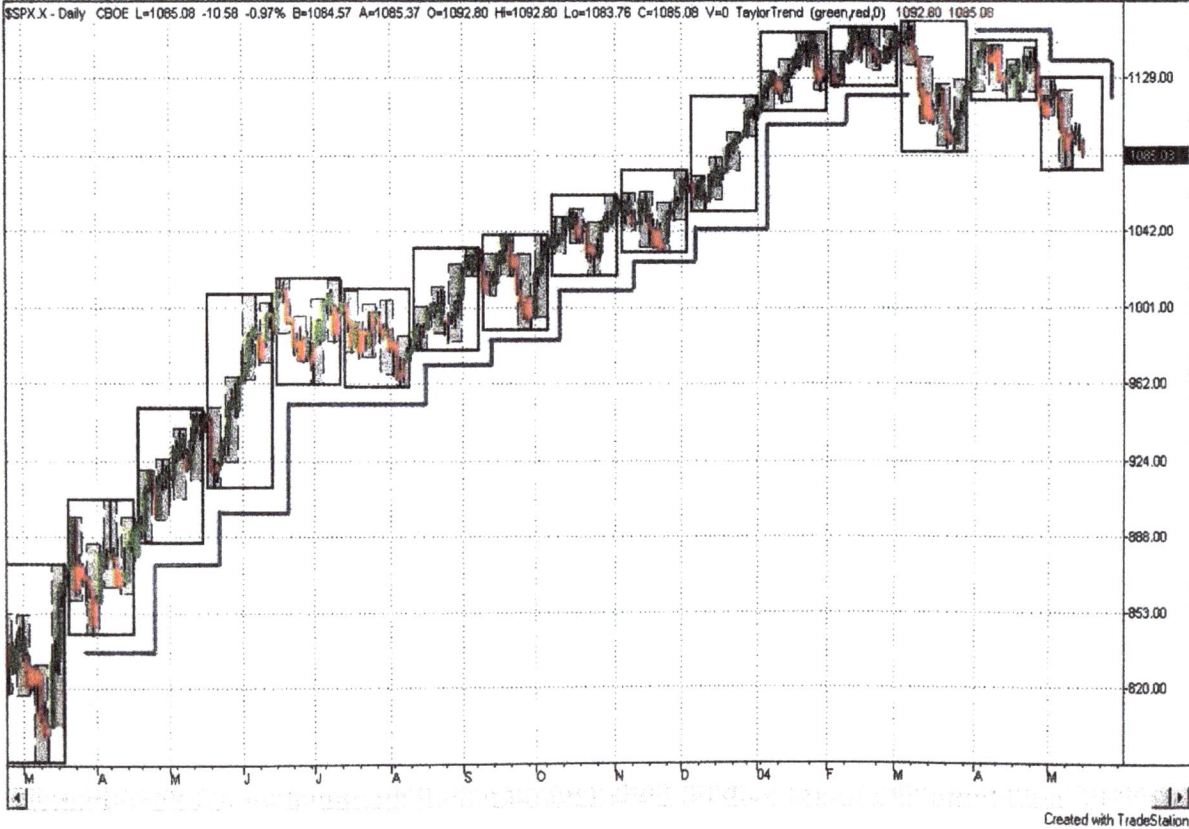

Note how clear the "stair step" pattern is on each monthly bar group.

USING INSIDE BARS TO ENTER IN THE DIRECTION OF THE LARGER TREND

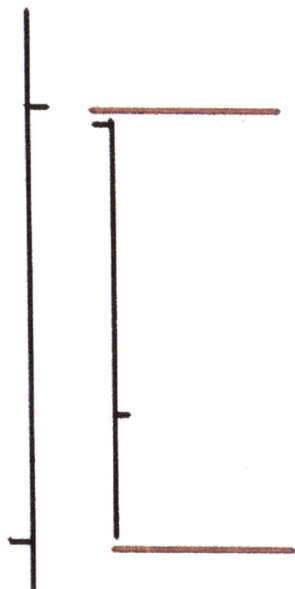

Inside Bars are contained within the range of the prior bar. In other words, it will have a high that is less than or equal to the prior high and a low that is greater than or equal to the prior low.

Inside bars actually have a reliably good track record in identifying trend changes in a wide variety of markets and time frames. Typically if the market forms an inside bar, the high and low of that bar will identify the future trend direction. If the market closes above the high of the inside bar, the trend is up. If the market closes below the low of the inside bar, the trend is down. Combining this tendency with our multi-perspective trend information can provide very nice low risk, high reward trading opportunities. If our longer-term trend is "UP," we only take buying signals. If the longer-term trend is "DOWN," we only take selling signals, just as we did with all of the other techniques. Remember the basic rule is to only trade in the direction of the larger trends. If the trend is broken, we look for double and triple top or bottom formations to trade.

TRADING THE INSIDE BAR

If we have the market in an "UP" trend, and an inside bar fauns, we would enter long as soon as the market closes above the high of the inside bar. If we have the market in a "DOWN" trend, and an inside bar forms, we would enter a short position as soon as the market closes below the low of the inside bar. Please note that inside bars can form on weekly or monthly bar groups as well. The closing bar of the group determines the entry signal. The chart on the following page illustrates how these patterns look on our multi-perspective charts.

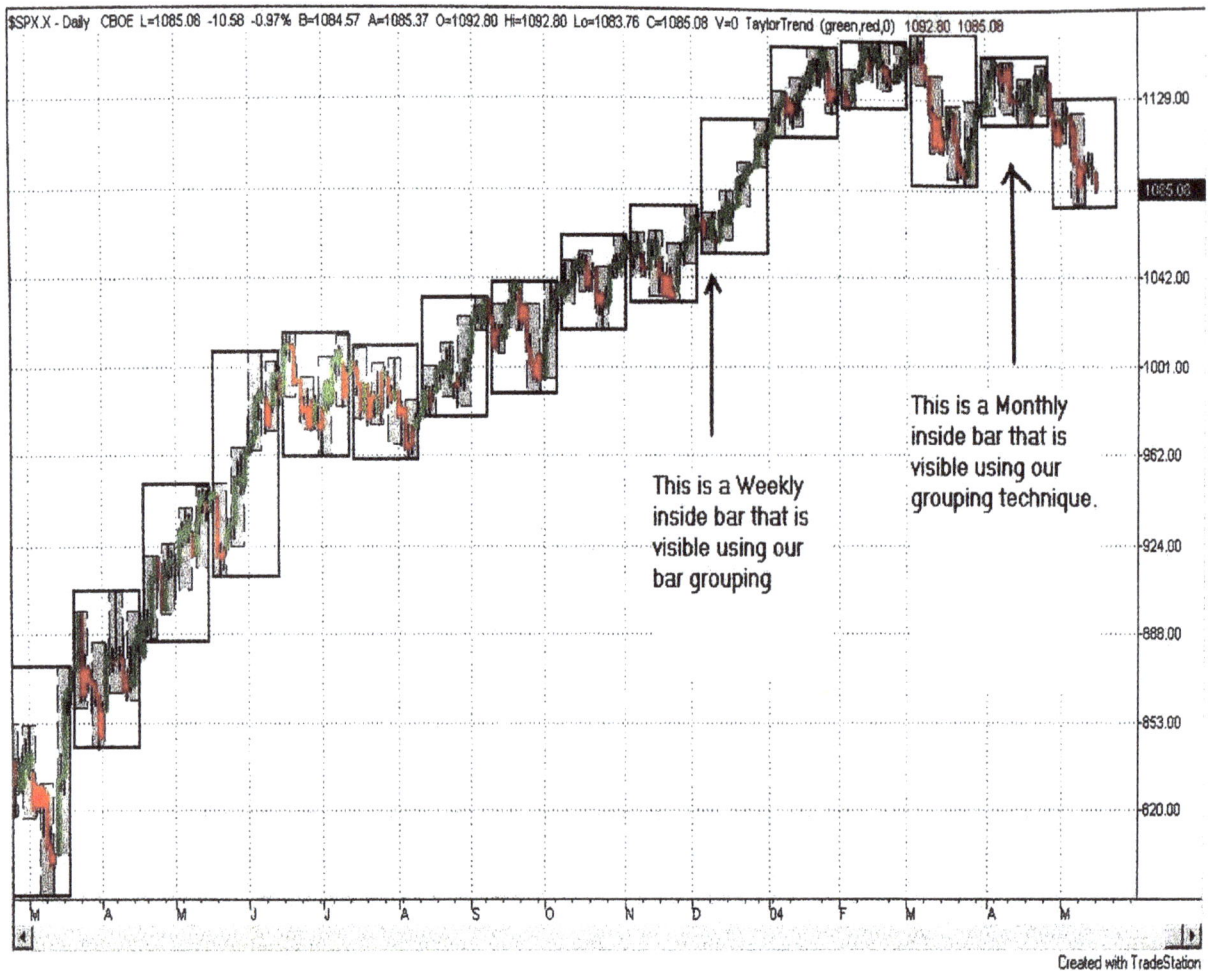

The next chart will show some of the daily inside bars that occurred during the uptrend. Notice that once the trend was broken, no additional signals were permitted.

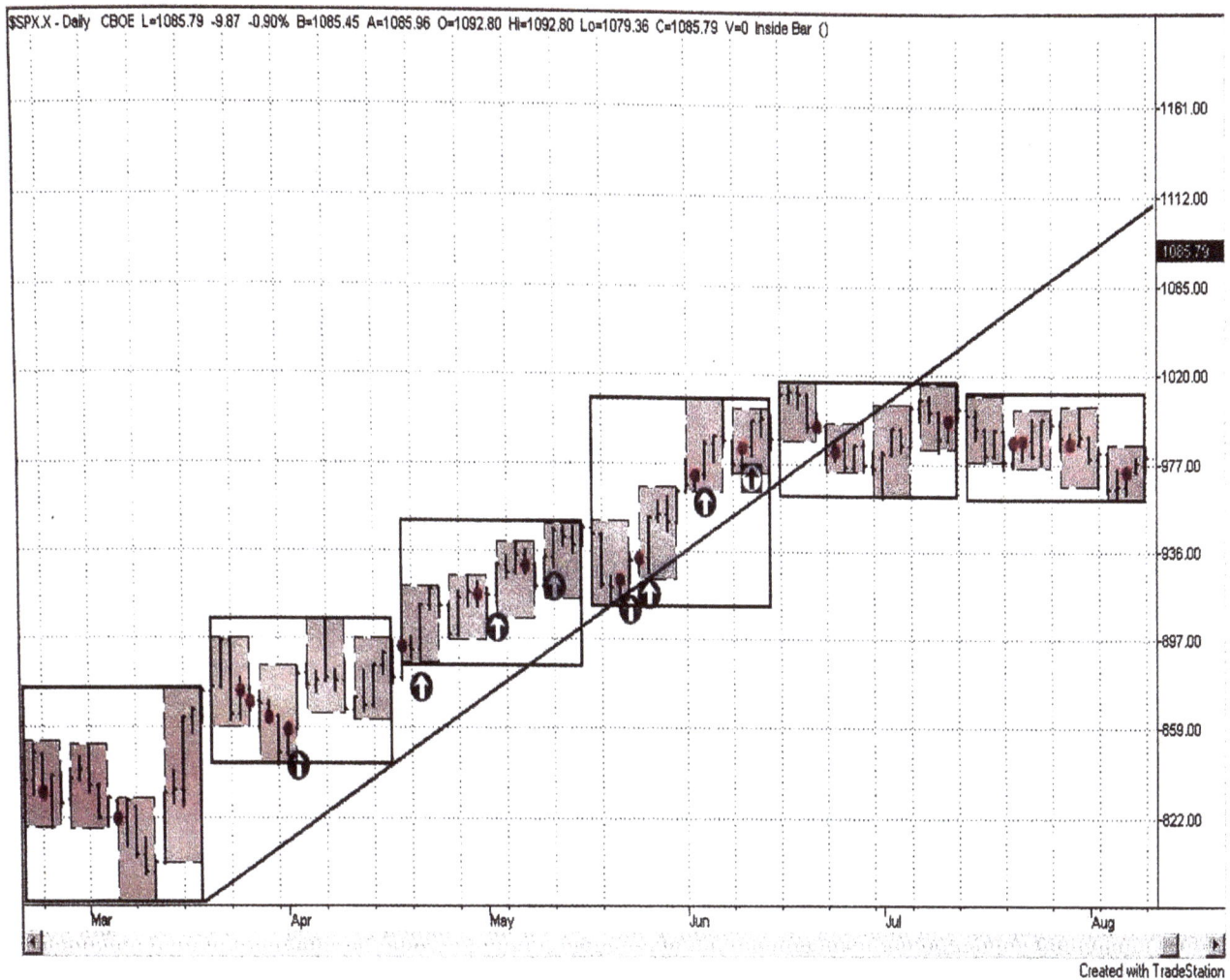

$SPX.X - Daily   CBOE L=1085.79 -9.87 -0.90% B=1085.45 A=1085.96 O=1092.80 Hi=1092.80 Lo=1079.36 C=1085.79 V=0 Inside Bar ()

Created with TradeStation

This is a zoomed in view of the chart and trend line that was shown on page 50. The chart will show a zoomed in view of the 50% retracement area illustrating the short sell entry signals from the daily inside bars. I turned off the paintbar study, so that the bars would stand out clearly. In addition, I had TradeStation mark each daily inside bar with a dot so that they stood out for easy reference and comparison.

61

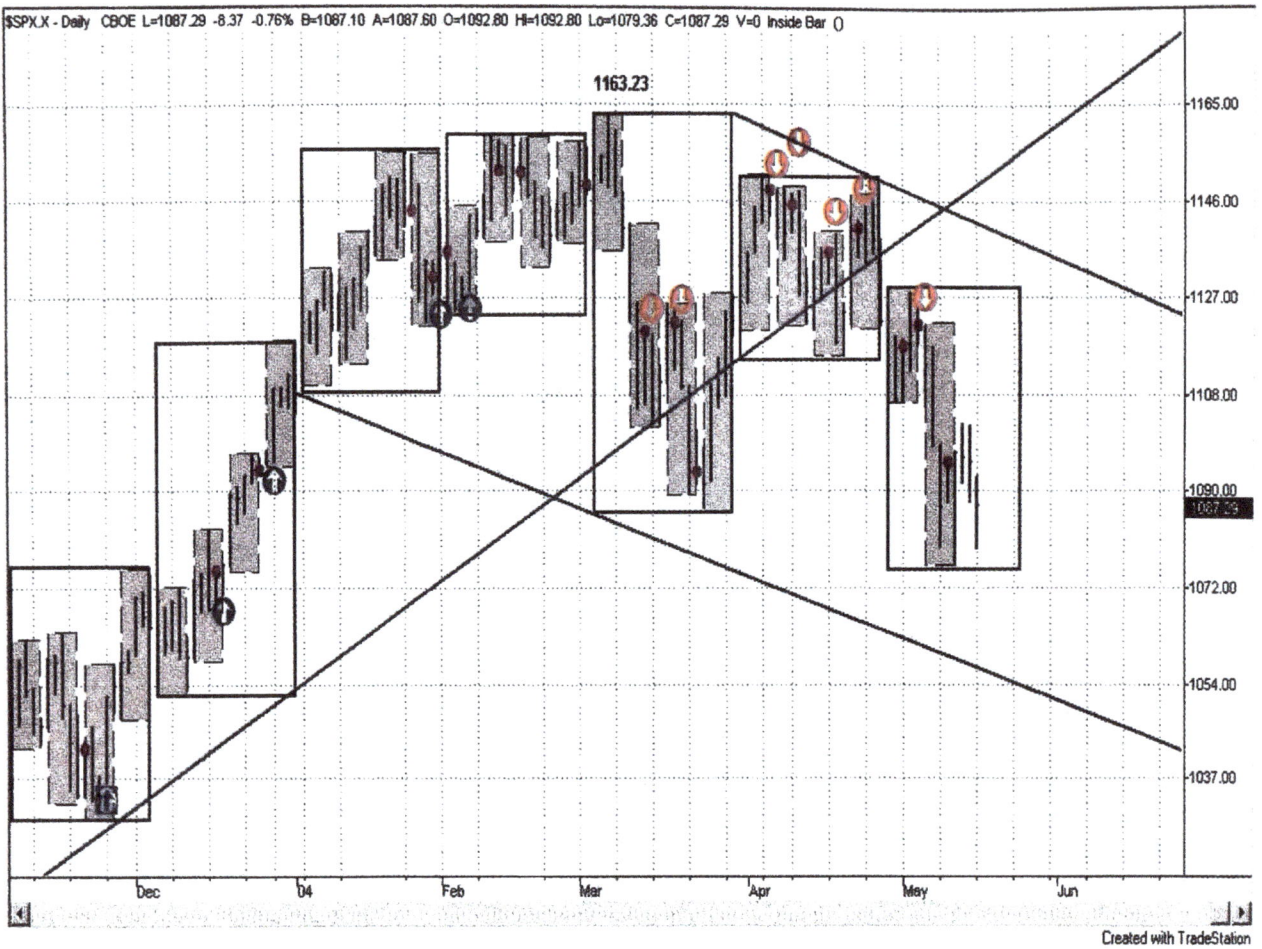

This technique can only be used in the direction of the larger trends. If you are kicked out of a trade, inside bars usually provide a number of opportunities to get back into the market. I showed these examples last, because they are my least favorite of the entry approaches that have already been shown. Study these examples and make sure that you are comfortable with the approach. Also, try to find examples in your favorite markets to gauge its usefulness.

UNDERSTANDING THE OPTIONS OPPORTUNITY

It is my opinion that every trader should learn at least a few simple options strategies. Options are the one investment vehicle that offers the greatest leverage and upside reward potential with a limited amount of risk. Sure, you can still lose money, but that is the case with all financial vehicles. A major advantage to options is their flexibility and versatility. They can be conservative or very speculative depending on the rules of your strategy. Options allow you to customize your position to your specific set of circumstances. Consider some of the following benefits of options:

- **You can protect a long position in the market from a decline in prices.**
- **You can protect a short position in the market from an advance in prices.**
- **You can prepare to buy a stock or commodity at a lower price.**
- **You can prepare to sell a stock or commodity at a higher price.**
- **You can position yourself for a big market move even when you don't know which way prices will move.**
- **You can benefit from a price rise or decline without incurring the cost of buying the stock outright or posting the minimum margin requirements.**

HOW OPTIONS BASICALLY WORK

Much like stocks and commodities, options can be used to take a long or short position in a market as a means to capitalize on an upward or downward price move. However, unlike stocks or futures, options can provide a trader or investor the benefits of leverage over a position in an individual stock, commodity, or even an index reflecting the broader market. Option buyers can also take advantage of predetermined, limited risk. However, the converse is also true, those who write options to collect a premium assume potentially unlimited risk if they do not manage or hedge their positions.

What is an option? An option is defined as the right to buy or sell a specific stock, commodity or other security for a specific price on or before a specific date but not the obligation. In other words, you are never obligated to buy or sell the underlying stock or commodity. A call option gives you the right to buy, while a put option gives you the right to sell. I used to remember this by thinking you call someone UP on the phone and if you don't like something, you put it DOWN. The person who purchases an option regardless of whether it is a put or a call is the option buyer. The person who sells the option (put or call) is known as the option seller or the option writer.

Options are like a type of contract in which all the terms are known. The terms of the contract are standardized and gives the option buyer the right, but not the obligation, to buy or sell a specified asset (like a stock or commodity) at a fixed price (strike price) for a specific amount of time, known commonly as expiration. To the buyer, a call option typically represents the right to buy 100 shares of a specified stock or 1 contract of a commodity. A put option usually represents the right to sell 100 shares of a specified stock or 1 contract of a commodity. The seller or writer of an option is obligated to perform according to the terms of the options contract. For example, selling the stock or commodity at the contracted price (the strike price) for a call seller, or purchasing it for a put seller in the event the option is exercised by the option buyer. The price paid for an option is called the premium. The potential loss to anyone that purchases options cannot be greater than the premium paid for the option, regardless of how the underlying asset performs. This specific feature of options allows a trader or investor to control the amount of risk assumed on the position. The option seller or writer assumes the risk of being assigned to fulfill the contract if it is exercised in return for the premium that was received from the option buyer. In strict accordance with standardized terms, all options have a limited time feature and expire on a certain date, called the "expiration date." The majority of options expire on the third Friday of the month they are listed for. For example, an option that is listed to expire May 2004 will typically expire on the third Friday of that month. So most of the options that have May 2004 expiration dates would expire on Friday the 21st. There are also long term options on stocks that can have expirations dates up to three years out (LEAPS). Most traditional options will not go out further than 9 months to expiration. There are primarily two styles of options (1) American-Style options, which are the most common and most frequently traded and (2) European-Style options, which have slightly different regulations in regards to expiration and the exercising rights of the option owner. An American-Style option is an option contract that may be exercised at any time between the date of purchase and the expiration date. European-Style options are different and can only be exercised during a specified time period just prior to expiring. Most European-Style options are primarily cash settled. Regardless of the type and differences in exercising rights, the option owner can sell both European and American Style options at any time prior to the expiration date. Therefore, if you have a profit in an option position there is absolutely nothing preventing you from cashing in.

CALL OPTIONS: The purchaser of a call option is paying for the right to buy the underlying stock or commodity at the stated strike price. If you were anticipating an advance in prices, you would be interested in the call options available for your market.

PUT OPTIONS: The purchaser of a put option is paying for the right to sell the underlying stock or commodity at the stated strike price. If you were anticipating a decline in prices, you would be interested in the put options available for your market.

The simplest strategy for using options is if you anticipate a certain directional trend in the price of a stock or commodity. Options will give you the right to buy or sell that market at a predetermined price, within a limited time window and can provide a very attractive risk reward opportunity due to the enormous leverage that they can provide. The decision to purchase a put option or a call option is dependent on whether you expect the price to go up or go down. If you anticipate higher prices, then purchasing a call option opens up the opportunity to participate in the upside potential of the specific market you are following without having to risk more than a small fraction of its market value. If you anticipate a downward move in the stock or commodity, purchasing put options will provide the opportunity to position yourself to benefit from a decline in price with the risk limited to the premium you paid to own it. The outright purchase of options offers you the ability to position yourself in accordance with your price trend expectations in a manner that allows you unlimited profit potential that is protected with an absolute limited risk. You can never lose more than the premium you paid to own the option. You can of course lose 100% of this premium, but it is typically not a large investment at all. Options can allow you to participate in price movements without committing large amounts of capital that would be required to buy or sell shares of a stock outright or post margin on a commodity futures contract.

What affects the price of an option? There are many factors that can affect the current value or strike price of an option. The main factors are the options strike price in comparison to the current price of the market, the amount of time left to expiration, volatility, interest rates, and dividends (for stock options). The most important consideration is the current value of the underlying market compared with the strike price of the option. This comparison determines if the option is in the money or out of the money. If your option is in the money in means that the current price is above your strike price if you own a call option and/or the current price is below your strike price if you own a put option. This means that your option has intrinsic value. For example, if you owned a call option with a strike price of 50 and the current price of the market is 55, then your option is in the money by 5 points. If you own a put option with a strike price of 35 and the current price of the underlying market is 20, then your option is in the money by 15 points. In other words, it has this much value because you can exercise it and immediately sell short the underlying market at 35. If your option is out of the money then the only value it really has is time. The amount of historical volatility will also affect the premium price that is charged to have time. For example, if you own a call option with a strike price of 50, and the price of the underlying market is 45, your option is out of the money by 5 points. Since it is out of the money, it has absolutely no intrinsic value. You or anyone else for that matter would not exercise your right to purchase something at 50 when the current market is selling it for 45. The only value it has is based on the amount of time until expiration. You are paying for the "chance" so to speak. If the underlying stock or commodity has been very volatile, then the option writer will charge a higher premium to protect his level of risk based on price move probabilities. The longer the amount of time left to expiration, the higher the premium paid for time value. The time value premium of options decays very rapidly during the last month to expiration. Options that are out of the money and have a month or less until expiration are usually very cheap. I have purchased out of the money options with only a few days of time value for 1/8 and sold them for $1, $2, and even $5. This is a huge return on the money. I would strongly suggest that you follow the day-to-day options prices (puts and calls) for the market of your choice so that you can appreciate how much these things can fluctuate in value. It is not uncommon for the price to fluctuate 50% to 300% in a day. Selective option buying (puts and calls) is a great way for a new trader to get started in a highly leveraged market. The cost of entry is relatively low and the returns can be almost unbelievable at times. As an added bonus, the limited risk feature of option buying ensures that a new trader will not get in over his head in the markets. When you decide to set up an option-buying program, you want to concentrate on the following criteria and rules:

• Follow the exact same buy and sell signals outlined previously in the course. Purchase call options for "buys" and put options for "short sells."

• Only purchase options that have strike prices that are "out-of-the-money." This means call options that have strike prices above the current market price and put options that have strike prices below the current market price.

• Only purchase options that have LESS than 3 months to their expiration date or maturity. Options with 30 days or less to expiration can be extremely inexpensive, but the time value will decay rapidly. However, I suggest that you watch how trading these options would work out with the swing trading strategies presented. These short-term options can provide enormous leverage because the option writer is basing his premium on the probabilities of time and volatility.

• Do not hold on to your options position until expiration. When the market price has achieved 75% of the anticipated price move, take profits on the options immediately. For example, if you have a profit target that would require the market to move 40 points, you would close out your options position after the market moved in your favor (40 x 0.75) = 30 points. You can also base your profit targets on simple returns. If your option doubles or triples in value, sell it or hedge it. This will be made clearer as we progress.

• Based on the prior rule, we can now refine our "out-of-the-money" purchases. We want to purchase strikes that are currently "out-of-the-money" but we anticipate that they will be "in-the-money" if we get our price move. To calculate this strike price, we just take our profit target or anticipated price move and multiply it by 37.5% and then add or subtract this amount from the current price. This gives us the basic area we should be shopping for the option's strike prices. For example, the current market is at 100 and we are expecting a decline of 20 points to our first support area. We would be looking for out of the money put options with strike prices 7.5 points (20 x 0.375) below the current price of 100. This gives us strikes around 92 or 93 for our puts.

• Do not hold onto losing options! They will be worthless at expiration unless they go into the money and have intrinsic value. If your options are not showing a profit within two weeks (1012 trading days from purchase), just close them out at a loss at the current market premium price. Just like any trading technique, a percentage of your option-buying trades will lose money. The name of the game is using favorable risk to reward opportunities and strict money management.

• Do not exercise your options to take a position in the stock or commodity. This involves added commission fees that are unnecessary expenses to your trading. Options commissions are cheap. You are better off simply selling your options prior to expiration at a profit or loss.

• Always use limit orders when purchasing or selling options. Just because a market maker has to sell you an option, doesn't mean he has to sell it to you at a price that will allow you to make money. The options market is typically not as liquid as the underlying market and fills can often be bad using market orders. If your limit order doesn't get filled within 10 to 15 minutes, change it based on the current "bid" and "ask" price for your particular strike price. Adjust your limit order up in price by $^1/_{8th}$ at a time.

- Markets go up and down, so make sure that you take advantage by purchasing the appropriate call or put option. In other words, don't be afraid to play both sides of the market.

- To gain additional leverage, re-invest 50% of your profits into your option purchasing program. Over time, this will allow you to increase the number of markets you are following and making trades in. Look at everything, stocks, indexes, currencies and commodities. With a little experience, you can just look at a chart and instantly see the swing pattern setups and long-term trends.

Remember that as the option buyer, you have the luxury of deciding when and where you will get out prior to expiration. There will be no margin calls with purchasing options. You simply buy them and the price + commission paid is your maximum risk. Take some time each day to study the options markets and follow the above rules and you will see the tremendous opportunity. You should also purchase an options book to learn additional options trading strategies. Options offer tremendous flexibility and will also allow you to be pretty creative with your trading positions. There are hedging strategies for longer-term positions, several types of spread trades, covered option-writing strategies for income, etc. Going into all of these details is beyond the purpose of this book, and there are plenty of really good options books on the market today. You already have what they don't, a method of anticipating tops and bottoms in price movement and the underlying trend. Profit targets are easily determined by looking at your support or resistance levels.

PROBABILITY OF A PRICE MOVE

This is a nice trick I learned and borrowed from the options guys. The option writer is basically trying to sell options each day that have a probability of being worthless at expiration. As the premium decays, they buy them back at a lower price to close out their position and remove the unlimited risk exposure they have on the trade. When you write options, you are basically taking a short position in the options market. Option writers are "selling short" the premium. In other words, you are expecting the premium to decay, allowing you to buy it back at a lower price, or letting it go all the way to zero. Option writers use simple formulas to predict the probabilities of a price move. It is kind of like an insurance underwriter. They want to collect premiums, not pay out on the policy. The option writer will use this probability information to write options that are likely to expire worthless. One of the easiest and most useful formulas is the following:

PRICE CHANGE = current price x historical volatility x square root of days left to expiration—all divided by the square root of the number of trading days in a year.

Trading days in a year is used as a constant number (252). The square root of this is 15.875. **H**

ISTORICAL VOLATILITY = (52-week high — 52-week low) / (52-week high + 52-week low)/2.

A quick example should get you up and running. Let's take a hypothetical stock XYZ. We look at the chart over the past year (52 weeks) and find the highest high and the lowest low within this time period. Do not go back further than 1 year. Let's assume that XYZ had a 52-week high of $125 and a 52-week low of $83. This gives us a historical volatility of:

(125 — 83) / (125 + 83)/2 or simply (42) / (104), which = 0.404 for historical volatility.

Now, let's assume that the current price of XYZ = $90 and we are looking at options that will all expire in the next 30 days. We just plug these numbers into the price change formula and get:

PRICE CHANGE = $90 x 0.404 x Sqrt(30) / 15.875. Calculating the square root of 30, this gives us ($90 x 0.404 x 5.48) / 15.875 = 12.55 for our price change calculation. This means that there is approximately a 70% chance that the market (XYZ in this example) will stay within +/- $12.55 of its current price. This means that the 102.5 (90 +12.55) call options and the 77.50 (90 —12.55) put options have a 70% chance of being worthless by expiration. The greater the price move from this price change of 12.55, the greater the odds are that the option will expire completely worthless. For example, if you double the 12.55 to $25, you will increase the probability to 95%. In other words, the $115 call option (90 +25) and the $65 put option (90 —25) have a 95% chance of being worthless in the next 30 days (expiration date). If you multiply the price change by 1.5, you get approximately 80% probabilities. Multiplying by 1.75 will give around 87% probabilities. This formula is very useful if you decide to write covered calls or covered puts (also known as married calls and puts) for premium income on longer-term investments. It is also useful to plug into our options strike price formula. For example, we might be able to purchase an inexpensive "out-of-the-money" option with only 3 weeks until expiration that is likely to be "in-the-money" based on our anticipated swing trading price target. We can plug in the above simple formula to find out what the option writer believes about the anticipated price move. If the odds are 65% or greater in his favor, the option is likely to be cheap and we can benefit from an extremely leveraged opportunity. As I have said before, I have seen options go from being worthless (1/16th) all the way up to being worth $5 in one or two days! If you follow sound money management principles, it does not take many winners to generate a positive financial outcome with this kind of risk to reward ratio. Put in the study time daily and you will see what I'm talking about. This is not intended to be a course in options trading, so I suggest that you purchase a book specifically targeted to options. Authors David Caplan and Larry McMillan would be my recommendation for options trading material.

UNDERSTANDING OPTION SPREADS

Spread strategies are one of the most popular trading techniques used by investors and traders in the options markets. Spread trades are popular for two reasons. They are typically lower in risk than buying or selling short the underlying market outright. They are flexible and highly adaptive to a variety of market conditions. There are many types of spreads that can be created, so you will most likely always be able to find one that suits your current needs. Don't overlook spread trading because you feel that it is too complicated. Spread trading is really pretty simple after you learn the techniques involved. Once you understand the advantages and disadvantages of spread trading and compare the risks with the potential rewards, you will be happy to have this technique in your trading toolbox. Spreads are designed primarily to limit risk, and in most instances, they perform this function very effectively as long as you, the trader or investor, knows exactly what you are doing!

WHAT ARE SPREADS? All spread trades have certain common characteristics. All spreads require you to simultaneously purchase and write options upon opening the trade. The options purchased versus the options written will have different strike prices, different expirations, or a combination of the two. Therefore, when you initiate a spread trade, you are simultaneously purchasing and writing options. When you "lift" or close a spread trade, you simultaneously buy back the options you wrote and sell the options you purchased. For example, let's say the market (XYZ) is at $90 and we feel it is going up to $100. We could purchase the $94 or $95 call (10-pts x 0.375) and write the $100 call. The premium received would offset some of the cost paid for the $94 call reducing our risk capital in the trade. Ideally, we would close out the spread when the market hits (10 x .75) $97.50 or 75% of our anticipated price move. This would mean that the option we purchased is now "in-the-money" and the one we wrote is still "out-of-the-money." This would be a good time to close the spread, also known as "lifting the spread." This particular type of spread is commonly known as a "vertical spread." This type of spread always consists of an option purchase and a further "out of the money" option write with the exact same expiration date. For example, a bearish vertical spread in the XYZ example would consist of purchasing an $86 put option and simultaneously writing the $80 put, receiving the premium to reduce the total cost of the trade.

There are many types of spread trades that you should be aware of Primarily, they fall into two categories (1) vertical spreads and (2) horizontal spreads. Basically, vertical spreads are based on price spreads, like the example above, and horizontal spreads are based on time to expiration. The time spread, which is also known as a calendar spread, is the simultaneous purchase of an option (put or call) of one expiration and the sale or writing of an identical put or call with a different expiration month but same strike price. Typically, you purchase the option with the greater time value (longer expiration date) and write the option with the shorter time value or expiration date. You are basically trying to capture the more rapid premium decay of the shorter-term option. For example, let's say it is February and you purchase the June 95 call and write the March 95 call simultaneously, you have entered a horizontal time spread or calendar spread. If you do this same technique, but use different strikes, you create what is known as a diagonal spread. This is sort of a blend between the vertical " price" spread and the horizontal "time" spread. For example, if you purchase the June 95 call and write the March 100 call, you create a diagonal spread. With expiration rapidly approaching, the rate of premium decay in the March 100 call will erode the price rapidly while the June 95 call holds more steadily due to its larger time value and better strike price.

THE BUTTERFLY SPREAD is simply a combination of two vertical spreads. For example, if we did both of the vertical spreads in the XYZ example all at once, we would have a butterfly spread. In other words, we would have purchased the 95 call and sold the 100 call and also purchased the 85 put and sold the 80 put at the same time. Some option traders call this same technique a sandwich spread. When executed correctly, this type of spread offers a good return on a fairly broad range of prices.

69

Many times this strategy can pay off when the market makes the "two lower swing tops or the two higher swing lows pattern" Also watch the performance of this spread at triple tops and triple bottoms on the swing chart.

DELAYED VERTICAL SPREAD OR "FREE TRADE" - The free trade is simply a delayed vertical spread. For example, let's say we purchase the 95 calls for 2½ ($250) and the market moves up and we can now sell the 100 calls for $2^1/_2$ points or more. This position is now free of charge and we can potentially make the $5 spread (100 — 95) on the trade with absolutely zero risk at this point.

THE COST FREE HEDGE- This options trade is typically based on having a position in the underlying market and writing a covered option and using the premium received to purchase an opposing directional option. For example, let's say we are short the soybean market and we have good profits in the trade and we want to protect these profits from a potential advance. We could write covered puts against our short position and purchase call options with the premium we receive. This way, if the market goes up, we make money on the calls, while the premium decays on the puts we sold. This, in effect, hedges our profits on the short position. This can also be done entirely with options. For example, let's say that you purchased the XYZ 90 call option and the market is now at $100 and you think that there is going to be a short term decline but you want to hold on to your long options position without losing too much of the current profits. You can sell or write the 105 call and purchase the 95 put. If the market explodes to the upside, you can still make at least the 5 point spread on the covered write. If the market declines, you own a put option that will increase in value as prices drop, resulting in a hedge.

THE STRADDLE- This options position is used when you anticipate a big price move, but don't know which direction the market will go. Here you purchase a put and a call with the same strike price and expiration date. The two combined premiums is your maximum risk in the trade. Therefore, you must expect prices to move greater than this cost. If you paid 6 points for the total position, the market must move at least this much one way or the other before you start making any real profits.

The main advantages to option purchasing and spread trading are limited and predetermined amounts of risk with a great deal of opportunity and leverage. Regardless of your choice to use them or not, it is worthwhile information to have in your overall trading knowledge. It provides an opportunity to trade many markets that might otherwise be financially out of reach. Take your time and learn as much as you can about this extremely flexible trading vehicle. As your confidence and trading surplus account grows, consider initiating an options trading program in a variety of markets and apply all of the new trading rules and methods you have learned to this new venture.

Required Options Discloser: Options involve risk and are not suitable for all investors. Prior to buying or selling options, a person must receive a copy of Characteristics and Risks of Standardized Options. Copies of this document are available from your brokerage firm or the Chicago Board Options Exchange (CBOE), 400 S. LaSalle Street, Chicago, IL 60605.

TIMING IMPORTANT STOCK MARKET BOTTOMS

The following is a simple buy only system for stock indexes or correlated investments, that I call "opportunist." It requires the daily close of the S&P500, NYSE advancing issues and declining issues. Each day, subtract the declining issues from the advancing issues (advance — decline). We will call the difference net breadth (advance — decline = net breadth). Construct an 18-period moving average of net breadth. This is a simple moving average, which is just the past 18-days added together and divided by 18 to produce an average eighteen-day value. When this average net breadth is **negative 400 or lower,** a buy signal is possible if other filters are in place.

SWING FILTER: The S&P500 must be down at least 14% from its last swing high in order to validate a buy signal. As an example, let's say the market makes an all time high at 1000. It declines 20% to 800 and our 18-day moving average of net breadth is less than negative 400. We will buy long as soon as the market can post an up close, i.e., today's close is greater than yesterdays close. Now, let's say the market goes up 16% to 928. This becomes our new swing high. If the market falls 14% or more from 928 and the moving average of net breadth falls below negative 400, we will get another buy signal.

CLOSE FILTER: The market must post a positive close after a 14% or greater decline in the S&P, and the average net breadth has fallen below negative 400.

The chart on the following page illustrates the signals with red up arrows for buys. Obviously, a trailing stop or exit strategy must be developed on your part. Options are another good way to trade these signals. Typically, the market will advance at least 14% as a minimum. The buy signal after the mini crash in 1990 is one exception to the rule. It only advanced 4.5% from the signal.

$SPX.X - Daily   CBOE L=1038.71 +4.93 +0.48%   O=1033.78   C=1038.64

337.78 400.00 -400.00

Created with TradeStation

The next 3 charts show other buy signals, which have occurred in the past 20 years.

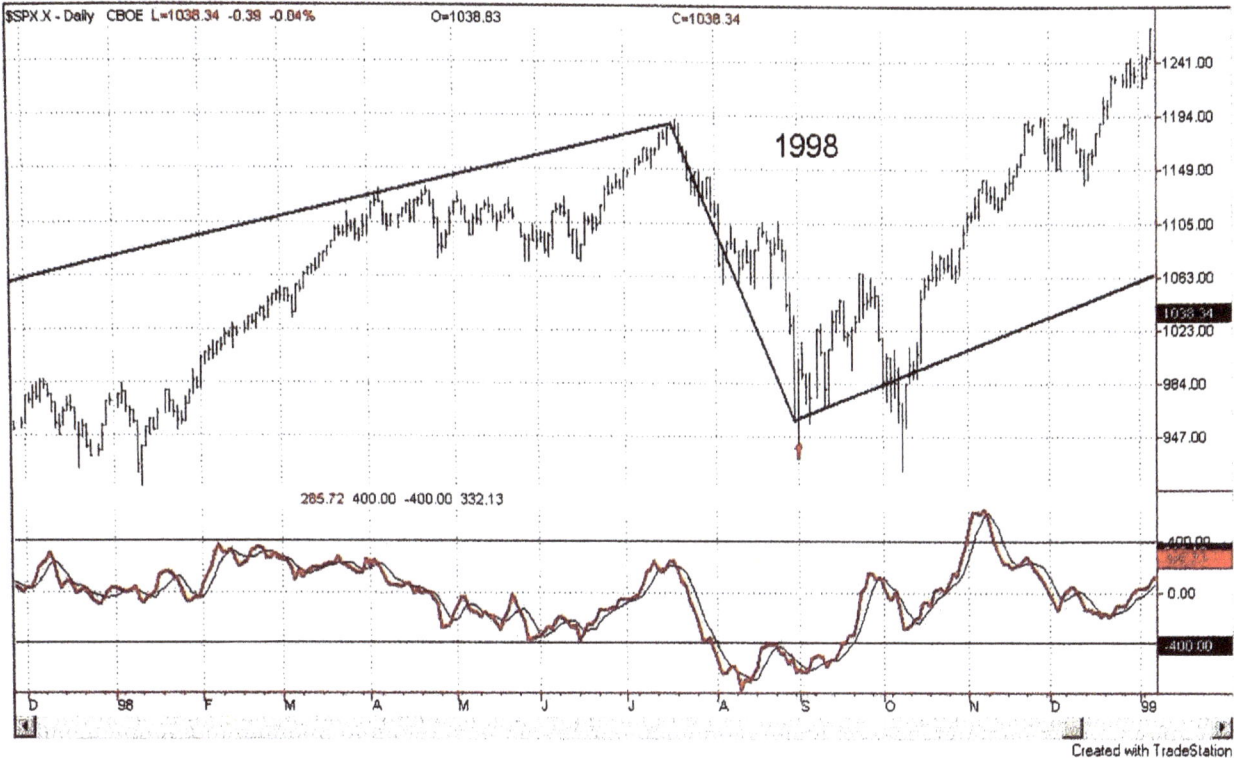

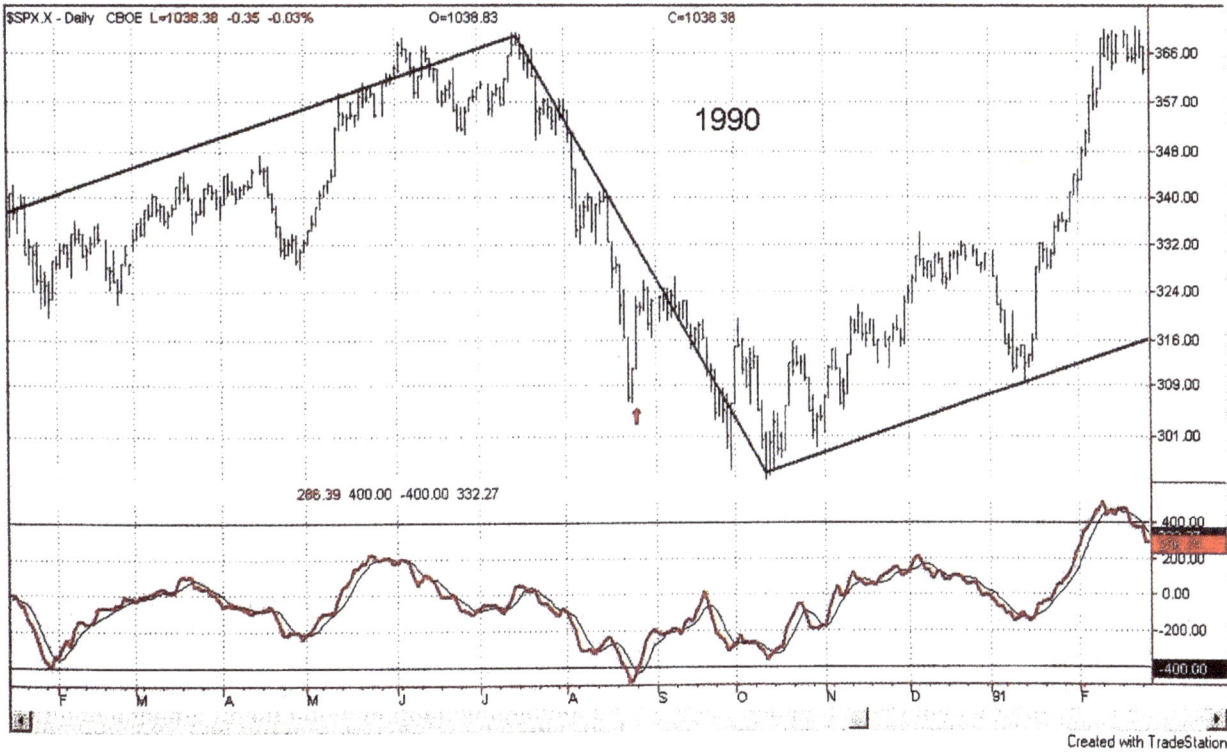

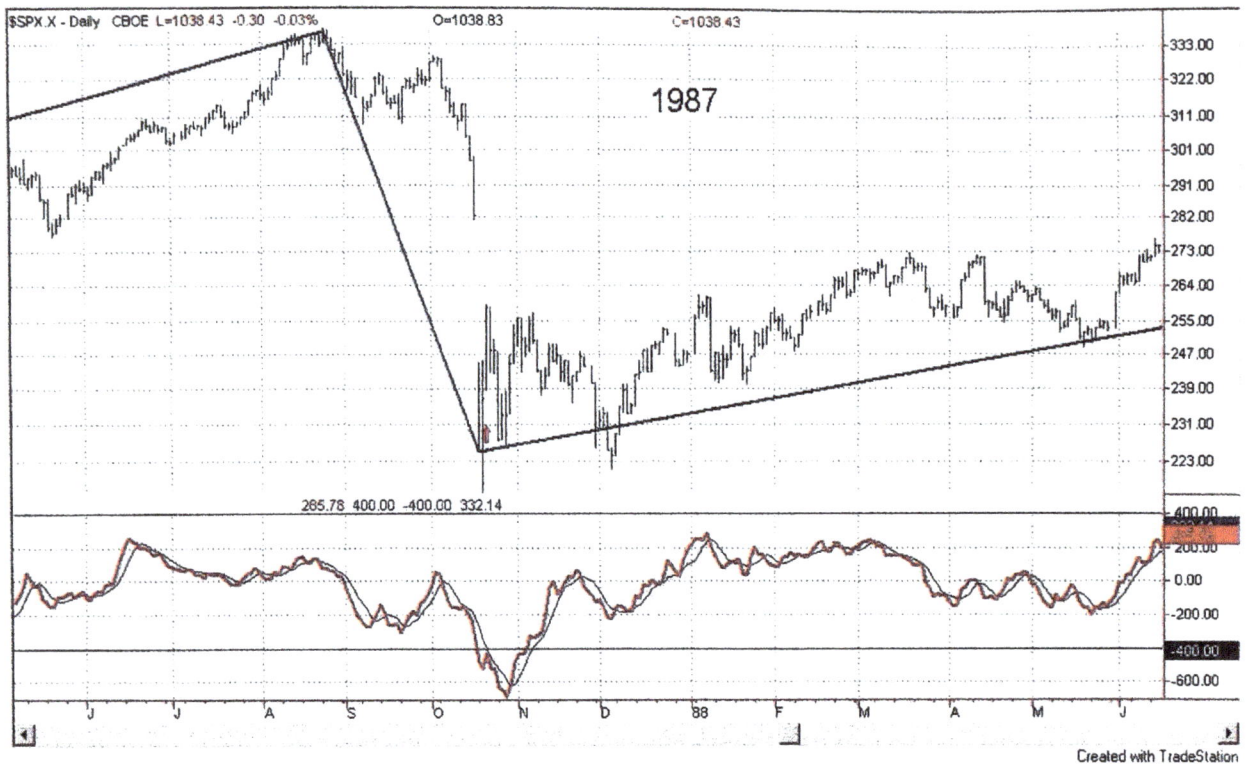

Created with TradeStation

THE IMPORTANT RE-ENTRY RULE!

In the event that you get stopped out of your position, which could have potentially happened in 1987, 1990, 1998, and possibly even the buy signal on 2/14/2003, depending on your risk management approach, you can re-enter long, i.e., buy again as soon as the market closes above the lowest close that occurred in the prior buy setup. A few examples will make this perfectly clear. The charts on the following pages illustrate this technique. Also, please notice the bullish divergence between a comparison of the S&P500 and the advance-decline line. The stock market makes new lows prior to the re-entry signals, but the advance-decline line is showing an up-trend compared to the last low.

Notice the bullish divergence at the re-entry point.

Here again, the market makes a new low, but the advance-decline line is showing a bullish up-trend.

If you are a stock trader, a mutual fund investor, or just like trading the NASDAQ-100 or S&P500 indexes, this system has provided me with many great trading opportunities over the years. Combining this simple system with the bar grouping and swing trading techniques is also advised to define exit strategies or potential profit targets. When the system gives a "buy" signal, there are several ways you can take advantage of the signal.

(1)   You can purchase call options on the SPX, OEX or NDX.

(2)   You can purchase tracking stocks like SPY, QQQ, DIA.

(3)   You can purchase call options on stocks that are leaders in these 3 major Indexes (Dow Jones Industrial Avg., NASDAQ-100 or S&P500), such as MSFT, INTC, IBM, GE to name a few.

(4)   You can purchase stocks outright.

(5)   You can go long in the futures market using the S&P futures or E-mini, the NASDAQ futures and/or mini contract as well as the Dow Industrial Futures Contract.

(6)   You can purchase aggressive stock mutual funds or index funds. Typically, options or futures will provide the biggest bang for your buck, but as I have said before, you must fully understand these trading vehicles before you dive into the markets.

SUCCESSFUL TRADING

It should be more than clear to you by now that successful trading is not exactly what you originally thought it was. Accuracy is not the deciding factor to consistently making money in the markets. Relying on the trading advice from newsletter writers and market gurus will not help you. Understanding Fibonacci and other magical numbers and price retracements is meaningless without money management and strict trading rules. Learning entry techniques without money management is about as valuable as flipping a coin. In reality, it is actually worse. Understanding Gann, Elliott Wave, or any other past legend may help a little, but it is not the "Holy Grail." Gann actually tried to point out the path with his "Twenty-Four Never Failing Rules," which are all money management rules. However, most people think he made his fortune entirely based on having amazing accuracy. I believe that he made his money by following strict rules and nothing more. Sure, he has made some amazing predictions and forecasts, but so have I. Trading is a completely different animal entirely. I'm not condemning any approach to the markets at all. Cycles work, astrology works, Elliott Wave works, Gann methods work, trend line breaks work, technical analysis works, fundamental analysis works, moving averages work, oscillators work, chart patterns work, etc. However, no approach, no matter how good, will work without sound money management rules because no approach is 100% infallible. These are the exact rules that most traders completely ignore and are the main reason why 80 to 90% of them fail. They believe that if they can find a way to predict the next top or the next bottom, they will turn everything around and become successful in the markets. Achieving any degree of success in this regard will eventually lead them to failure because they are basing their entire success on being correct instead of controlling risk. Pulling off a few winning trades will reward you for the wrong reasons and conditions you to behave incorrectly in all future trades.

There are hundreds of books on trading, and all types of different approaches, systems and methods, yet they produce so few successful traders. Money management is not the main attraction to the markets and it probably never will be, but ironically, making money is. How will you ever make money in the markets without "money management" skills? Making money only requires you to protect enough of your capital so that you will always be able to have enough to continue playing the game. If you can consistently take limited risks with high reward potential, you don't have to be correct very often to be profitable. That's what trading really is! Learn this fact NOW and you will find yourself well ahead of the majority of traders.

WHAT IS LUCK?

My favorite defmition of luck is this: *Luck is when opportunity meets preparation.* In the financial markets, opportunities come more frequently than in any other business or profession, but you must have the required knowledge and experience in order to benefit from them. You can create your own luck simply by being prepared for opportunity to present itself. Bad luck results when opportunity either passes you by because you did not recognize it or you get in over your head because you are tempted to take action even though you are completely unprepared and lack the necessary skills to prevail in the venture. Some so-called lucky people win a fortune in the lottery or some other chance venture. Consider this surprising statistic: most "lucky" lottery winners end up broke, exactly back to where they started. I just listened to an interview the other week, where this guy shared in one of the largest lotto winnings of all time. He received a little over $300,000 per year for 20 years and now he is broke and the payments have stopped. Sure, he lived it up for 20 years, but now he has nothing. He was not prepared for the enormous opportunity. Just because you receive a windfall doesn't mean it will last forever. You must always keep a level head and make wise choices. Remember SUCCESS has "U" in it. To achieve success requires "U" to follow necessary rules and steps. I have shared as much as possible in this regard. It is now up to you to make sure that you are prepared when opportunity presents itself. This book has now come full circle, completing the cycle initiated at the beginning. I could provide additional charts and examples to bulk things up, but I have found from teaching others, that this actually tends to confuse people rather than help them. It is much better to provide a few clear examples than it is to fill up a book with hundreds of charts that do not serve the greater purpose of instructing.

Opportunity

They do me wrong who say I come no more,
When once I knock and fail to find you in;
For every day I stand outside your door
And bid you wake and rise to fight and win.

Wail not for precious chances passed away,
Weep not for golden ages on the wane!
Each night I burn the records of the day;
At sunrise every soul is born again.

Laugh like a boy at splendors that have fled,
To vanished joys be blind and deaf and dumb;
My judgments seal the dead past with its dead,
But never bind a moment yet to come.

Though deep in mire wring not your hands and weep;
I lend my arm to all who say, "I CAN!"
No shame-faced outcast ever sank so deep
But yet might rise and be again a man!

Dost thou behold thy lost youth all aghast?
Dost reel from righteous retribution's blow?
Then turn from blotted archives of the past
And find the future's pages white as snow.

Art thou a mourner? Rise thee from thy spell,
Art thou a sinner? Sins may be forgiven,
Each morning gives thee wings to flee from hell,
Each night a star to guide thy feet to heaven.

—Walter Malone

This concludes the trading section of the course. As I have stated throughout the material, these swing-charting methods can and should be analyzed on multiple time frames, especially daily, weekly and monthly charts just as Gann did. That does not mean that you have to trade on each and every signal, but it should be used to keep you informed of the potential or significance of a specific price level. You are encouraged to maintain these levels on your daily, weekly and monthly swing charts! It is also my hope that you are drawing at least some of your charts by hand. I have found many huge double tops and bottoms that took years to form that provided unbelievable opportunities.

The only way to see these opportunities is to have longer-term charts where you can carry the old price levels across the page as I have instructed you to do time and time again. The markets repeat and react in a similar fashion around old price extremes because people "REMEMBER" buying or selling at specific prices. My grandfather is 87 years old and he still remembers exactly when he made purchases in Ford Motor Company stock and how much it was trading for when he bought it and he hasn't owned any stocks at all since 1986. People react to specific prices simply because of memory. Maintaining longer-term charts with the knowledge you have gained in this course of instruction will prove truly invaluable. Options provide one means to expose yourself to these longer-term patterns with limited risk, but you can also trade them by other means. I should also mention that there are other markets that should be brought to your attention. One is FOREX currency trading and the other is stock futures contracts. Opportunities are everywhere, when you have sound money management and a method to exploit an advantage. The principles in this course should work on any market because the basic advantage we are exploiting exists in all markets and time frames. When you have the risk to reward equation stacked in your favor, you can lose much more often than you win and still profit tremendously. This is the true secret to winning in this business.

Please protect this information and keep it to yourself. I have much more knowledge to share, but will only do so if my rights are not violated. Copies of some of my other materials are being pirated due to irresponsible persons. You paid for this information and not someone else, so please keep it to yourself. Anyone interested in programming the techniques described in this course is required to get my written approval first.

A MINI-COURSE IN THE SCIENTIFIC METHODS OF W.D. GANN

GANN'S CUBE

♈ ❻ ♌ ♎ ♐ ♒

0, 60, 120, 180, 240, 300

WHAT GANN SAID ABOUT "THE HEXAGON CHART"

Since everything moves in a circle and nothing moves in straight lines, this chart is to show you how the angles influence stocks at very low levels and very high levels and why stocks move faster the higher they get, because they have moved out to where the distance between the angles of 45° are so far apart that there is nothing to stop them and their moves are naturally rapid up and down.

We begin with a circle of "1," yet the circle is 360° just the same. We then place a circle of circles around this circle and six circles complete the second circle, making a gain of 6 over the first one, ending the second circle at 7, making 7 on this angle a very important month, year, and week as well as day, the seventh day being sacred and a day of rest. The third circle is completed at 19. The fourth circle around is completed at 37, a gain of 18 over the previous circle. The fifth circle is completed at 61, a gain of 24 over the previous circle. The sixth circle is completed at 91, a gain of 30 over the previous circle and the seventh circle at 127, a gain of 36 over the last circle. Note that from the first circle, the gain is 6 each time we go around. In other words, when we have traveled six times around the hexagon, we have gained 36 or 6x6. Note that this completes the first hexagon and as this equals 127 months, shows why some campaigns will run 10 years and seven months, or until they reach a square of the hexagon, or the important last angle of 45°.

The eighth circle around is completed at 169, a gain of 42 over the first. This is a very important angle and an important time factor for more reasons than one. It is 14 years and one month, or double our cycle of 7 years. Important tops and bottoms culminate at this angle, as you will see by going over your charts.

The ninth circle is completed at 217, a gain of 48 over the previous circle. The tenth circle is completed at 271, a gain of 54. Note that 271 is the ninth circle from the first, or is the third 90° angle or 270°, three fourths of a circle, a strong point... All this is confirmed by the Master Twelve Chart, by the four seasons and by the Square of Nine chart, and also confirmed by the Hexagon Chart, showing that mathematical proof is always exact no matter how many ways or from what directions you figure it.

The eleventh circle is completed at 331, a gain of 60 over the last circle. The twelfth circle is completed at 397, which completes the hexagon, making a gain in 11 circles of 66 from the beginning. 66 months, or 5 years and six months, marks the culmination of major campaigns in stocks. Note how often they culminate on the 60th month, then have a reaction, and make a second top or bottom in the 66th month. Note the number 66 on the master twelve chart. Note it on the Square of Nine and note that 66 occurs on an angle of 180° on the hexagon chart, all of which confirms the strong angle at this point.

We have an angle of 66°, one of 67.5° and one of 68, confirming this point to be doubly strong for tops and bottoms or space movements up or down.

Note the number 360° on the hexagon chart. It completes a circle of 360°. From our beginning point this occurs at an angle of 180° on the hexagon chart going around, but measuring from the center, it would equal an angle of 90° or 180°, making this a strong point, hard and difficult to pass, and the ending of one campaign and the beginning of another.

Again with the center of the hexagon chart at "1" notice that 7, 19, 37, 61, 91, 127, 169, 217, 271, 331, and 397 are all on this direct angle and are important points in time measurement.

Beginning with "1" and following the other angle, note that 2, 9, 22, 41, 66, 97, 134, 177, 226, 281, and 342 are all on a same angle of 90°, or an angle of 60° and 240° as measured by the hexagon chart.

Go over this chart and the important angles each way and you will see why the resistance is met on days, weeks, months, or years, and why stocks stop and make tops and bottoms at these strong important points according to time.

When any stock has passed above 120° or especially above 127° or 127 points and gone out of the square of the first hexagon, its fluctuations will be more rapid and it will move faster up and down. Notice near the center that in traveling from 66 to 7 you strike the angle of 180° or 90°, but when the stock gets out to 162, it can travel up to 169 before striking another strong angle. That is why fast moves occur up and down as stocks get higher and as they move from a center of time.

Remember that everything seeks the center of gravity and important tops and bottoms are formed according to centers and measurements of time from a center, base, or beginning point, either top or bottom. The angles formed going straight up and across, 'may a com' just the same going across as the stock travels over days, weeks, months, or years. Thus, a stock going up to 22.5 would strike an angle of 22.5°. If it moved over 22.5 days it would strike the angle or 22.5 weeks or 22.5 months, it would also strike an angle of 22.5°, and the higher it is when those angles are struck and the angle it hits going up, the greater the resistance to be met. Reverse that rule going down.

Market movements are made just the same as any other thing; they are constructed. It is the same as constructing a building. First, the foundation has to be laid and then the four walls have to be completed, and last, but not least, the top has to be put on. The cube or hexagon proves exactly the law, which works because of time and space in the market. When a building is put up, it is built according to a square or hexagon. It has four walls or four sides, a bottom and a top; therefore, it is a cube.

In working out the 20-year cycle in the stock market, the first 60° or 5 years from the beginning forms the bottom of the cube. The second 60°, running to 120°, completes the first angle or the first side and runs out the 10-year cycle. The third 60° or the second side ends 15 years or 180°. It is very important because we have the building half completed and must meet the strongest resistance at this point. The fourth 60°, or at the end of 20 years or 240 months, completes the third side. We are now two thirds around the building, a very strong point that culminates and completes our 20-year cycle. The fifth 60°, or 300° point, days, weeks, or months completes 25 years, a repetition of the first five years, but it completes the fourth side of our building and is a very important angle. The sixth 60°, or 360°, completing the circle and ending 30 years as measured by our time factor, which runs 1° per month on an angle of 45°, completes the top. This is a complete cube and we begin over again.

Study this in connection with the hexagon chart. It will help you.

JAN 1931
W.D. GANN

The time factor that moves 1° per month is heliocentric Saturn. Each 60° of longitude takes an average of 5 years to complete. However, because of the elliptical orbits of planets, this time period is not equal to an exact 5 calendar years, which is based upon the Earth's orbit around the Sun. For example, if you add 180° of longitude to heliocentric Saturn's position from the August 1982 low in the U.S Stock Market and also add 360° of longitude from the Stock Market top on December 1St, 1968, you arrive at April 22, 1998. If you look at a chart of the market averages from this date you will see that this marked a significant top where the majority of all stock averages dropped 20% or more. If you add 60° of longitude to the 1982 low, you get the crash low of 1987, etc. Some of the other important time periods mentioned in Gann's description of the hexagon chart are 20 years, 14 years, 7 years, 45 years and $67^1/_2$ years. Here is a bonus tip: Heliocentric Mars also has a hexagon cycle that runs every 60° starting from 18° of Taurus.

20 years is the synodic period of Jupiter & Saturn, i.e., the planets conjoin every 20 years. 240° of heliocentric Saturn or 240 months. This is just four 60-degree hexagon turns.

14 years is the synodic period of Jupiter & Uranus and also is the average time required for Neptune to change signs or move 30° of longitude.

7 years is $^1/_2$ the synodic period of Jupiter & Uranus and is also the average time required for Uranus to change signs or move 30° of longitude.

45 years is the synodic period of Saturn & Uranus.

67% years is $1^1/_2$ times the synodic period of Saturn & Uranus or 360° + 180° = 540°.

See Gann's Hexagon Chart below and study all of the above in connection with the Hexagon Chart.

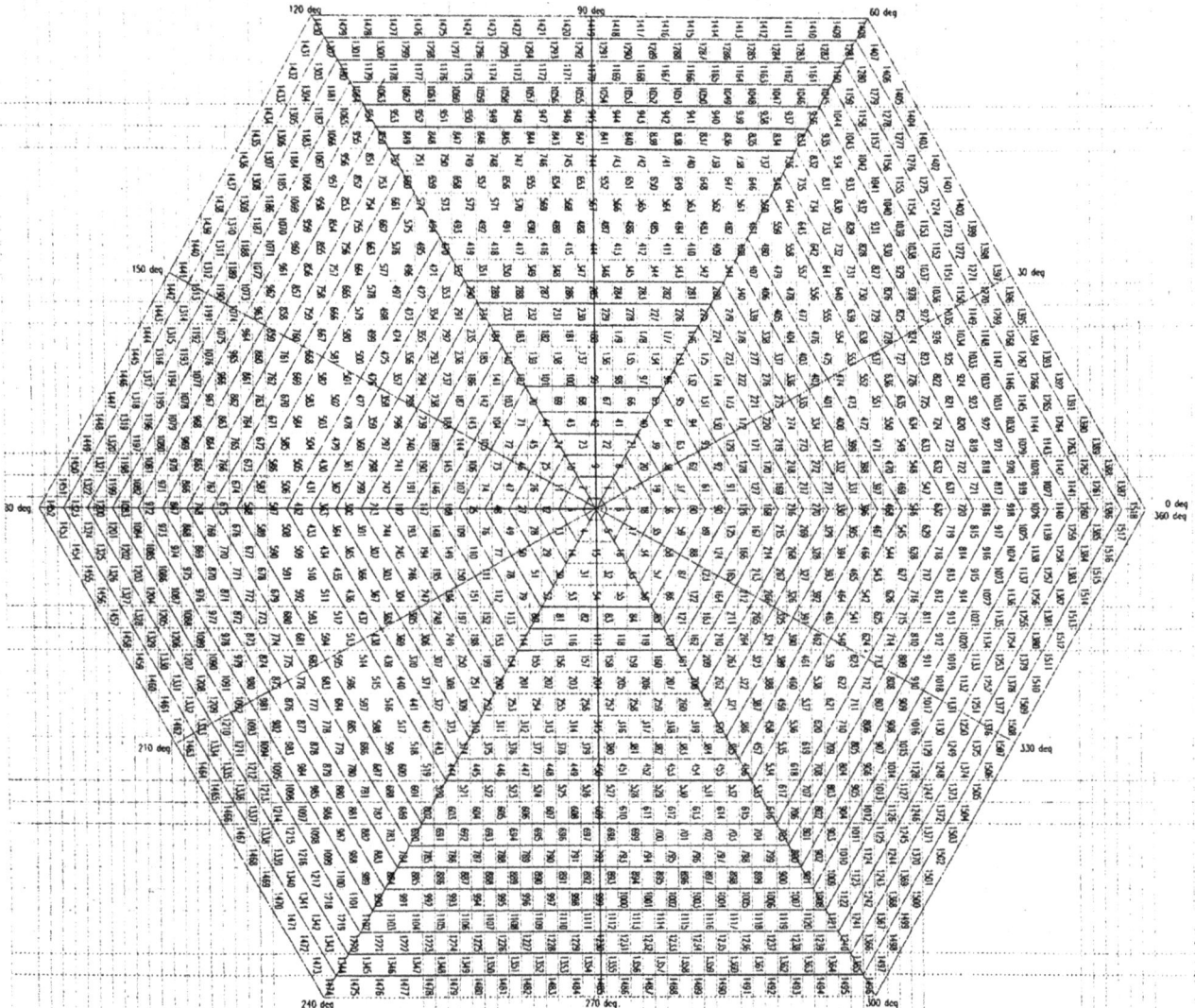

In Gann's Egg Course, there is a quote that pertains to measuring time periods, which pertains to the basic construction of the Square of Nine and hexagon charts. It reads:

"Man first learned to record and measure time by the use of the sundial, and by dividing the day into 24 hours of 15 degrees in longitude. The "reflection" of the geometrical angle on the sundial indicated the time of day. Since all time is measured by the sun, we must use the 360 degrees of the circle to measure time periods for the market, but remember, you must always begin to count time as days, weeks, and months from extreme high and extreme low levels, and not from exact seasonal or calendar time periods. 45 days is 1/8th of a year, 90 degrees is 1/4th of a calendar year, or a square. $112^{1}/_{2}$ days is $90 + 22^{1}/_{2}$. 120 is 1/3rd of the circle and is a triangle. 135 is $90 + 45$, 150 is $90 + 60$, $157^{1}/_{2}$ is $135 + 22^{1}/_{2}$,165 is $120 + 45$. 180 is $^{1}/_{2}$ of a complete circle or opposite to 0, the starting point. Very important for a change in trend. $202^{1}/_{2}$ is $180 + 22^{1}/_{2}$, 225, a 45-degree angle is $180 + 45.240$, a triangle is 2 times 120. $247^{1}/_{2}$ is $225 + 22^{1}/_{2}$. 270 is $^{3}/_{4}$ of a circle and 3 squares of 90. $292^{1}/_{2}$ is $270 + 22^{1}/_{2}$, 315 is $270 + 45$, 337 $^{1}/_{2}$ is $315 + 22^{1}/_{2}$ and 360 degrees is the complete circle. You measure weekly and monthly time periods in the same way as you do the days and watch all of these important time angles for a change in trend.

The reason for constructing a chart like the hexagon, or even the Square of Nine, is really based upon the hypothesis that each positive whole number, i.e., the regular counting numbers 1, 2, 3, 4, 5, etc., all correspond to some specific angle of a circle between 0° and 360°. By relating the counting numbers to degrees of a circle, Gann used astrological aspect theory to determine future levels of price support or resistance. In astrology, the 45-degree class of aspects (0, 45, 90, 135, 180, 225, 270, and 315) are considered a negative influence. Using this theory, these same degree relationships applied to price movement could define future price levels where a change in trend is possible.

Most traders misuse these anticipated price levels. If you calculate that there should be some resistance at the price level of $51.00, you should not expect that this calculation should work out to the exact tick. In addition, you need to let the stock trade through the number Ideally, the stock would penetrate the price by $1 or $2 reaching between $52 and $53. Then when it declines back to $51.00, it would be a short sell with a stop just above the high reached at the $52-$53 level. The inverse is true for support levels. The stock should fall a little below your calculation and bounce back up to confirm a buying level with a stop loss just a tick or two below the actual low price reached. Experiment with this idea on some charts and you will get the feel for it. These rules also apply to the Square of Nine.

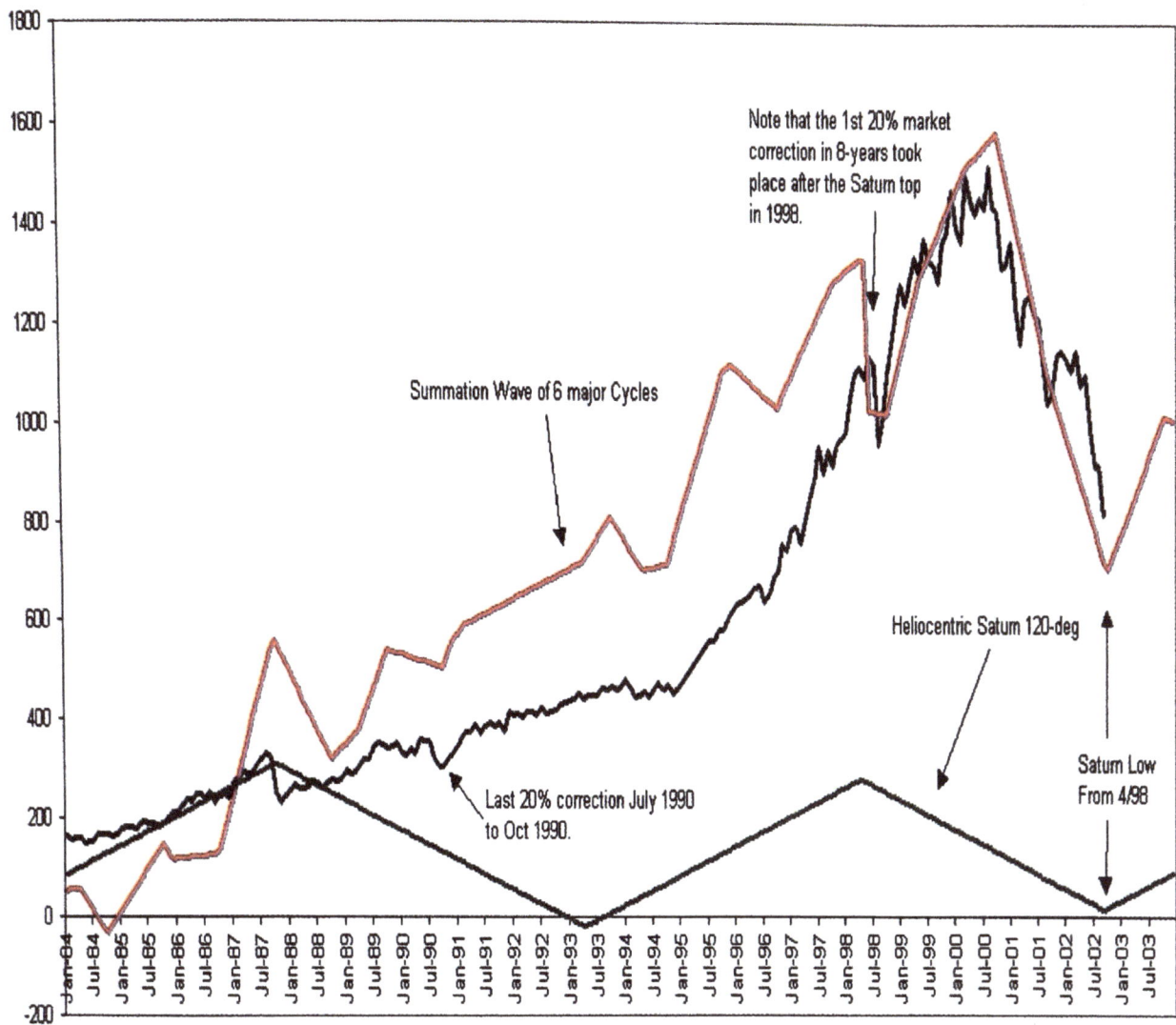

Stock Market Example of the Saturn Hexagon

Projected Saturn low is October 20, 2002. Venus transits 2-Aries on October 9th!

"Master Time Factor & Forecasting By Mathematical Rules"

++

W. D. GANN
88 WALL STREET
NEW YORK

Scientific Advice and Analytical Reports on Stocks and Commodities
Author of *Truth of the Stock Tape, Wall Street Stock Selector* and *The Tunnel Thru the Air*
Member American Economic Assn Royal Economic Society
Cable Address
"Ganwade New York"

FORECASTING

::::::::::::::::::::

Every movement in the market is the result of a natural law and a cause which exists long before the effect takes place and can be determined years in advance. The future is but a repetition of the past, as the Bible plainly states:

> *"The thing that hath been, it is that which shall be; and that which is done is that which shall be done, and there is no new things under the sun." Eccl. 1:9*

Everything moves in cycles as a result of the natural law of action and reaction. By a study of the past, I have discovered what cycles repeat in the future.

MAJOR TIME CYCLES

There must always be a major and a minor, a greater and a lesser, a positive and a negative. In order to be accurate in forecasting the future, you must know the major cycles. The most money is made when fast moves and extreme fluctuations occur at the end of major cycles.

I have experimented and compared past markets in order to locate the major and minor cycles and determine what years in the cycles repeat in the future. After years of research and practical tests, I have discovered that the following cycles are the most reliable to use:

GREAT CYCLE — MASTER TIME PERIOD — 60 YEARS:

This is the greatest and most important cycle of all, which repeats every 60 years or at the end of the third 20-year cycle. You will see the importance of this by referring to the war period from 1861 to 1869 and the panic following 1869; also 60 years later — 1921 to 1929 — the greatest bull market in history and the greatest panic in history followed. This proves the accuracy and value of this great time period.

50-YEAR CYCLE:

A major cycle occurs every 49 to 50 years. A period of "jubilee" years of extreme high or low prices, lasting from 5 to 7 years, occurs at the end of the 50-year cycle. "7" is a fatal number referred to many times in the Bible. It brings about contraction, depression and panic. Seven times "7" equals 49, which is shown as the fatal evil year, causing extreme fluctuations.

30-YEAR CYCLE:

The 30-year cycle is very important because it is one-half of the 60-year cycle or great cycle and contains three 10 year cycles. In making up an annual forecast of a stock, you should always make a comparison with the record 30 years back.

20-YEAR CYCLE:

One of the most important time cycles is the 20-year cycle or 240 months. Most stocks and the averages work closer to this cycle than to any other. Refer to analysis of the "20-year forecasting chart" given later.

15-YEAR CYCLE:

Fifteen years is three-fourths of a 20-year cycle and most important because it is 180 months or one-half of a circle.

10-YEAR CYCLE:

The next important major cycle is the 10-year cycle, which is one-half of the 20-year cycle and one-sixth of the 6-year cycle. It is also very important because it is 120 months or one-third of a circle. Fluctuations of the same nature occur which produce extreme high or low every 10 years. Stocks come out remarkably close on each even 10-year cycle.

7-YEAR CYCLE:

This cycle is 84 months. You should watch 7 years from any important top and bottom. 42 months or one-half of this cycle is very important. You will find many culminations around the 42nd month. 21 months or $1/4$ of this cycle is also important. The fact that some stocks make top or bottom 10 to 11 months from the previous top or bottom is due to the fact that this period is one-eighth of the 7-year cycle.

There is an 84-year cycle, which is 12 times the 7-year cycle, which is very important to watch. One-half of this cycle is 42 years — 1/4 is 21 years, and 1/8 is 10 1/2 years. This is one of the reasons for the period of nearly 11 years between the bottom of August 1921 and the bottom of July 1932. A variation of this kind often occurs at the end of a great cycle or 60 years. Bottoms and tops often come out on the angle of 135 degrees or around the 135th month or 11 $1/4$-year period from any important top or bottom.

5-YEAR CYCLE:

This cycle is very important because it is one-half of the 10-year cycle and $1/4$ of the 20-year cycle. The smallest compete cycle or workout in a market is 5 years.

MINOR CYCLES:

The minor cycles are 3 years and 6 years. The smallest cycle is one year, which often shows a change in the 10th or 11th month.

RULES FOR FUTURE CYCLES

Stocks move in 10-year cycles, which are worked out in 5-year cycles—a 5-year cycle up and a 5-year cycle down. Begin with extreme tops and extreme bottoms to figure all cycles, either major or minor.

RULE 1— A bull campaign generally runs 5 years-2 years up, 1 year down, and 2 years up, completing a 5-year cycle. The end of a 5-year campaign comes in the 59th or 60th months. Always watch for the change in the 59th month.

RULE 2 — A bear cycle often runs 5 years down — the first move is 2 years down, then 1 year up, and 2 years down, completing the 5-year downswing.

RULE 3 — Bull or bear campaigns seldom run more than 3 to 3½ years up or down without a move of 3 to 6 months or one year in the opposite direction, except at the end of major cycles, like 1869 and 1929. Many campaigns culminate in the 23rd month, not running out the full two years. Watch the weekly and monthly charts to determine whether the culmination will occur in the 23rd, 24th, 27th· or 30th month of the move or in extreme campaigns in the 34th to 35th or 41St to 42nd month.

RULE 4 — Adding 10 years to any top will give you the top of the next 10-year cycle, repeating about the same average fluctuations.

RULE 5 — Adding 10 years to any bottom will give you the bottom of the next 10-year cycle, repeating the same kind of year and about the same average fluctuations.

RULE 6 — Bear campaigns often run out in 7-year cycles, or 3 years and 4 years from any complete bottom. From any complete bottom of a cycle, first add 3 years to get the next bottom; then add 4 years to that bottom to get the bottom of the 7-year cycle. For example: 1914 bottom — add 3 years, gives 1917, low of panic; then add 4 years to 1917, gives 1921, low of another depression.

RULE 7 — To any final major or minor top, add 3 years to get the next top; then add 3 years to that top, which will give the third top; add 4 years to the third top to get the final top of a 10-year cycle. Sometimes a change in trend from any top occurs before the end of the regular time period; therefore you should begin to watch the 27th, 34th, and 42nd months for a reversal.

RULE 8 — Adding 5 years to any top will give the next bottom of a 5-year cycle. In order to get top of the next 5-year cycle, add 5 years to any bottom. For example: 1917 was the bottom of a big bear campaign; add 5 years gives 1922, top of a minor bull campaign. Why do I say; "Top of a minor bull campaign?" Because the major bull campaign was due to end in 1929.

1919 was top; adding 5 years to 1919 gives 1924 as a bottom of a 5-year bear cycle. Refer to Rules 1 and 2, which tell you that a bull or bear campaign seldom runs more that 2 to 3 years in the same direction. The bear campaign from 1919 was 2 years down-1920 and 1921; therefore, we only expect a one-year rally in 1922; then two years down — 1923 and 1924, which completes the 5-year bear cycle.

Looking back to 1913 and 1914, you will see that 1923 and 1924 must be bear years to complete the 10-year cycle from the bottoms of 1913-14. Then, note 1917 bottom of a bear year, adding 7 years gives 1924 also as a bottom of a bear cycle. Then, adding 5 years to 1924 gives 1929 top of a cycle.

FORECASTING MONTHLY MOVES

Monthly moves can be determined by the same rules as yearly:

Add three months to an important bottom, and then add 4, making seven, to get minor bottoms and reaction points.

In big upswings a reaction will often not last over two months, the third month being up, the same rule as in yearly cycle-2 down and the third up.

In extreme months, a reaction sometimes only lasts 2 or 3 weeks; then the advance is resumed. In this way, a market may continue up for 12 months without breaking a monthly bottom.

In a bull market, the minor trend may reverse and run down 3 to 4 months, then turn up and follow the main trend again.

In a bear market, the minor trend may run up 3 to 4 months, then reverse and follow the main trend, although, as a general rule, stocks never rally more than 2 months in a bear market; then start to break in the 3rd month and follow the main trend down.

FORECASTING WEEKLY MOVES

The weekly movement gives the next important minor change in trend, which may turn out to be a major change in trend.

In a bull market, a stock will often run down 2 to 3 weeks, possibly four, then reverse and follow the main trend again. As a rule, the trend will turn up in the middle of the third week and close higher at the end of the third week, the stock only moving 3 weeks against the main trend. In some cases the change in trend will not occur until the fourth week, then the reversal will come and the stock will close higher at the end of the forth week.

Reverse this rule in a bear market.

In rapid markets with big volume, a move will often run 6 to 7 weeks before a minor reversal in trend, and in some cases, like 1929, these fast moves last 13 to 15 weeks or $^1/_4$ of a year. These are culmination moves up or down.

As there are 7 days in a week and seven times seven equals 49 days or 7 weeks, this often marks an important turning point. Therefore, you should watch for a top or bottom around the 49th to 52nd day, although at times a change will start on the 42nd to 45th day, because a period of 45 days is 1/8 of a year. Also, watch for culmination at the end of 90 to 98 days.

After a market has declined 7 weeks, it may have 2 or 3 short weeks on the side and then turn up, which agrees with the monthly rule for a change in the third month.

Always watch the annual trend of a stock and consider whether it is in a bull or bear year. In a bull year, with the monthly chart showing up, there are many times that a stock will react 2 or 3 weeks, then rest 3 or 4 weeks, and then go into new territory and advance 6 to 7 weeks more.

After a stock makes top and reacts 2 to 3 weeks, it may then have a rally of 2 to 3 weeks without getting above the first top, then hold in a trading range for several weeks without crossing the highest top or breaking the lowest week of that range. In cases of this kind, you can buy near the low point or sell near the high point of that range and protect with a stop loss order 1 to 3 points away. However, a better plan would be to wait until the stock shows a definite trend before buying or selling; then buy the stock when it crosses the highest point or sell when it breaks the lowest point of that trading range.

FORECASTING DAILY MOVES

The daily movement gives the first minor change and conforms to the same rules as the weekly and monthly cycles, although it is only a minor part of them.

In fast markets, there will be a 2-day move in the opposite direction to the main trend and on the third day, the upward or downward course will be resumed in harmony with the main trend.

A daily movement may reverse trend and only run 7 to 10 days; then follow the main trend again.

During a month, natural changes in trends occur around

6th to 7th   14th to 15th 23rd to 24th

9th to 10th 19th to 20th 29th to 31st

Those minor moves occur in accordance with tops and bottoms of individual stocks.

It is very important to watch for a change in trend 30 days from the last top or bottom. Then watch for changes 60, 90, 120 days from tops or bottoms. 180 days or six months, this is very important and sometimes marks changes for greater moves. Also around the 270th and 330th day from important tops or bottoms, you should watch for important minor and often major changes.

January 2nd to 7th and 15th to 21st:

Watch these periods each year and note the high and low prices made. Until these high prices are crossed or low prices broken, consider the trend up or down.

Many times when stocks make low in the early part of January, this low will not be broken until the following July or August and sometimes not during the entire year. This same rule applies in bear markets or when the main trend is down. High prices made in the early part of January are often high for the entire year and are not crossed until after July or August. For example:

U. S. Steel on January 2, 1930 made a low at 166, which was the half-way point from 1921 to 1929, and again on January 7, 1930 declined to $167^1/_4$. When this level was broken, Steel indicated lower prices.

July 3rd to 7th and 20th to 27th:

The month of July, like January, is a month when most dividends are paid and investors usually buy stocks around the early part of the month. Watch those periods in July for tops or bottoms and a change in trend. Go back over the charts and see how many times changes have taken place in July, 180 days from January tops or bottoms. For example:

July 8, 1932 was low; July 17, 1933, high; and July 26, 1934 low of the market.

HOW TO DIVIDE THE YEARLY TIME PERIOD

Divide the year by 2 to get 6 months, the opposition point or 180-degree angle, which equals 26 weeks.

Divide the year by 4 to get the 3 months' period or 90 days or 90 degrees each, which is $^1/_4$ of a year or 13 weeks.

Divide the year by 3 to get the 4 months' period, the 120 degree angle, which is 1/3 of a year or 17-1/3 weeks.

Divide the year by 8, which gives $1^1/_2$ months, 45 days and equals the 45 degree angle. This is also $6^1/_2$ weeks, which shows why the 7th week is always so important.

Divide the year by 16, which gives 22 $\frac{1}{2}$ days or approximately 3 weeks. This accounts for market movements that only run 3 weeks up or down and then reverse. As a general rule, when any stock closes higher the 4th consecutive week, it will go higher. The 5th week is also very important for a change in trend and for fast moves up or down. The fifth is the day, week, month, or year of ascension and always marks fast moves up or down, according to the major cycle that is running out.

BULL AND BEAR CALENDAR YEARS

By studying the yearly high and low chart and going back over a long period of time, you will see the years in which bull markets culminate and the years in which bear markets begin and end.

Each decade or 10-year cycle, which is 1/10th of 100 years, marks an important campaign. The digits from 1 to 9 are important. All you have to learn is to count the digits on your fingers in order to ascertain what kind of a year the market is in.

No. 1 in a new decade is a year in which a bear market ends and a bull market begins. Look up 1901, 1911, 1921.

No. 2, or the second year, is a year of a minor bull market, or a rally in a bear market will start at some time. See 1902, 1912, 1922, 1932.

No. 3 starts a bear year, but the rally from the 2" year may run to March or April before culmination, or a decline from the 2nd year may run down and make bottom in February or March, like 1933. Look up 1903, 1913, 1923.

No. 4 or the 4th year, is a bear year, but ends the bear cycle and lays the foundation for a bull market. Compare 1904, 1914.

No. 5 or the 5th year, is the year of Ascension, and a very strong year for a bull market. See 1905, 1915, 1925, 1935.

No. 6 is a bull year, in which a bull campaign, which started in the 4th year, ends in the Fall of the year and a fast decline starts. See 1896, 1906, 1916, 1926.

No. 7 is a bear number and the 7t^h year is a bear year, because 84 months or $84\frac{3}{4}$ degrees is 7/8 of 90. See 1897, 1907, 1917, but note 1927 was the end of a 60-year cycle, so there was not much decline.

No. 8 is a bull year. Prices start advancing in the 7th year and reach the 90th month in the 8th year. This is very strong and a big advance usually takes place. Review 1898, 1908, 1918, 1928.

No. 9, the highest digit and the 9th year, is the strongest of all for the bull markets. Final bull campaigns culminate in this year after extreme advances and prices start to decline. Bear markets usually start in September to November at the end of the 9th year and a sharp decline takes place. See 1869, 1879, 1889, 1899, 1909, 1919, and 1929 - the year of the greatest advances, culminating in the fall of the year, followed by a sharp decline.

No. 10 is a bear year. A rally often runs until March and April; then a severe decline runs to November and December, when a new cycle begins and another rally starts. See 1910, 1920, 1930.

In referring to these numbers and these years, we mean the calendar years. To understand this, study 1891 to 1900, 1901 to 1910, 1911 to 1920, 1921 to 1930, and 1931 to 1939.

The 10-year cycle continues to repeat over and over, but the greatest advances and declines occur at the end of the 20-year and 30-year cycles, and again at the end of the 50-year and 60-year cycles, which are stronger than the others.

IMPORTANT POINTS TO REMEMBER IN FORECASTING

TIME is the most important factor of all and not until sufficient time has expired does any big move, up or down, start. The time factor will overbalance both space and volume. When time is up, space movement will start and big volume will begin, either up or down. At the end of any big movement— with monthly, weekly or daily—time must be allowed for accumulation or distribution.

Consider each individual stock and determine its trend from its position according to distance in time from bottom to top. Each stock works out its 1, 2, 3, 5, 7, 10, 15, 20, 30, 50, and 60-year cycles from its own base or bottoms and tops, regardless of the movements of other stocks, even those in the same group. Therefore, judge each stock individually and keep weekly and monthly charts on them.

Never decide that the main trend has changed one way or the other without consulting the angles from top to bottom and without considering the position of the market and cycle of each individual stock.

Always consider the annual forecast and whether the big time limit has run out or not before judging a reverse move. Do not fail to consider the indications on time, both from main tops and bottoms, also volume of sales and position on geometrical angles.

A **daily** chart gives the first short change, which may run for 7 to 10 days; the weekly chart gives the next important change in trend; and the monthly the strongest. Remember, **weekly moves** run 3 to 7 weeks, **monthly moves** 2 to 3 months or more, according to the yearly cycle, before reversing.

YEARLY BOTTOMS AND TOPS: It is important to note whether a stock is making higher or lower bottoms each year. For instance, if a stock has made a higher bottom each year for five years, then makes a lower bottom than the previous year, it is a sign of a reversal and may mark a long down cycle. The same rule applies when stocks are making lower tops for a number of years in a bear market.

When extreme advances or declines occur, the first time the market reverses over 1/4 to 1/2 of the distance covered in the previous movement, you consider that the trend has changed, at least temporarily.

It is important to watch **space movements.** When time is running out one way or the other, space movements will show a reversal by breaking back over 1/4, 1/3 or 1/2 of the distance of the last move from extreme low to extreme high, which indicates the main trend has changed.

Study all the **instructions and rules** that I have given you; read them over several times, as each time they will become clearer to you. Study the charts and work out the rules in actual practice as well as on past performance. In this way, you will make progress and will realize and appreciate the value of my method of forecasting.

HOW TO MAKE UP ANNUAL FORECASTS

I have stated before that the future is but a repetition of the past; therefore, to make up a forecast of the future, you must refer to the previous cycles.

The previous 10-year cycle and 20-year cycle have the most effect in the future, but in completing a forecast, it is best to have 30 years past record to check up, as important changes occur at the end of 30-year cycles. In making up my 1935 forecast on the general market, I checked the years 1905, 1915, and 1925. For the 1929 forecast, I compared 1919 —10 years back, 1909 — 20 years back, 1899 — 30 years back, and 1869 — 60 years back, the great cycle.

You should also watch 5, 7, 15, and 50 year periods to see if the market is repeating one of these closely.

MASTER 20-YEAR FORECASTING CHART
1831-1935

In order to make up an annual forecast, you must refer to my Master 20-Year Forecasting Chart and see how these cycles have worked out and repeated in the past.

As stated before, the 20-year cycle is the most important cycle for forecasting future market movements. It is one-third of the 60-year cycle and when three 20-year cycles run out, important bull and bear campaigns terminate.

In order for you to see and study how these cycles repeat, I have made up a chart of 20-year cycles, beginning with the year 1831. To show all of these cycles from 1831 to date, we have carried thru on this chart the monthly high and low on railroad and canal stocks from 1831 to 1855. Beginning with 1856, we have used the W. D. Gann Averages on railroad stocks until the beginning of the Dow-Jones Averages in 1896. After that, we used the Dow-Jones Industrial Stock Averages.

> After the end of the 20-year cycle in 1860,
> the next cycle begins at 1861 and runs to 1880,
>
> the next cycle begins at 1881 and runs to 1900,
>
> the next cycle begins at 1901 and runs to 1920,
>
> the next cycle begins at 1921 and runs to 1940.

By placing the monthly high and low prices for each of those 20-year periods above each other, it is easy to see how the cycles repeat. The year of the cycles are marked from "1" to "20." Study the chart and note what happened in the 8th and 9th year of each cycle —extreme high prices have always been reached. For example:

1929 FORECAST:

According to my discovery of the 60-year cycle, I figured that 1929 would repeat like 1869, 1909, and 1919. Looking back 20 years, we find that top was reached in August 1909, and 60 years before, top was reached in July 1869. If you will read my Annual Forecast for 1929, you will see that I had figured the top must come no later than the end of August and stated that a "Black Friday" would come in September. Strictly following the 1869 top, the top would have come in July, 1929, and some stocks did make top at that time. Following the 1909 top, we could expect top in August, and the actual high of the averages and many individual stocks was reached on September 3, 1929. Going back to 1919, we find that the averages made first top in July and a big decline followed, but extreme high was made in the early part of November.

From all of these tops — 1869, 1909, and 1919 — sharp declines followed in the fall of the year, just as they did in 1929. Therefore, you see how easy it was to follow this great advance and determine when it would culminate. There is no other way, outside of using the 20 and 60-year cycle that we could have forecast this great bull campaign and its culmination so closely in 1929. (See the Excel file called 1929WD-Gann.)

1869-73 VS. 1929-33:

After the 1869 top, stocks continued to decline and reached low in November 1873. See how many other bottoms were reached around this time in other cycles. After the big decline from 1929, notice that in October 1933 the last low was reached on the Dow-Jones Averages; then followed an advance to new high levels, crossing the top of July 1933.

1935 FORECAST:

Figuring out the forecast for 1935, we see on this 20-year chart that we are running against 1855, 1875, 1895, and 1915; therefore, we look to see what happened in those years. We find that in 1895 the high was reached in September; in 1915, the high of the year was reached in December.

Then, look back at 1865, 1885, 1905, and 1925, the years in the 5¹h zone or the 10-year cycles. We find that in 1865 the high was reached in October; in 1905, the high was in October; in 1925, the high was in November.

Then, we would have a good guide in making up the forecast for 1935 and would know what months to watch for top and a change in trend. My annual forecast for 1935, which was made up in October 1934, indicated top for October 28 and a secondary top for November 15-16, 1935.

There are other ways of using the chart to your advantage. One method of determining the trend is to compare the years of previous cycles in the same zone. For example, after the Dow-Jones 30 Industrial Averages crossed 108 in May 1935 they were above the average high price of all the previous years in the 15th year zone. Therefore, the market indicated higher prices and showed that there would be a bull campaign.

1936 FORECAST:

If we wish to make up a forecast for the year 1936, we compare the years in the 16th year zone, viz. 1856, 1876, 1896, and 1916. As 60 years back is a very important cycle, we look at 1876 first, then 1896, and 1916.

>1876 — We find that the averages run up and reach high in March, then decline to the end of the year.

1896 — Next, we look at 1896, which is 40 years back, or two 20-year cycles, a very important presidential election year, just as 1936 will be. We find that there was a moderate rally into February, a decline to March, then a small rally to May, from which a panicky decline followed, culminating on August 8, 1896, with the averages at the lowest levels in years. From that point a bull campaign started, with prices working higher to December.

1916 — The next important cycle is 20 years later, or 1916. We find that prices declined in January, rallied moderately in February, then declined sharply to April, rallied to June, then declined and made bottom in July, from which a big bull campaign started, making top in November 1916, in a war market. A panicky decline followed from the latter part of November into December.

This completes our comparison of the 60, 40, and 20-year cycles back from 1936. Next, we look up the cycles on the other side of the chart, in the 6th year of the 20-year cycle, or the 6th zone, or the 10-year cycles. These years are 1866, 1886, 1906, and 1926.

1866 — We find that in 1866 there was a sharp decline, reaching bottom in February; then an advance, with top of the year in October.

1886 — We find a sharp decline and bottom in January, a moderate rally into March, then a sharp decline in new lows in May; a sharp advance, reaching high in November, and a sharp decline in December.

1906 — The next important cycle to consider is 1906. In that year the great McKinley boom, which began in 1895, culminated. The railroad averages reached the highest price in history up to that time. From the high of January, a sharp decline followed to May. Much of this selling was caused by the San Francisco earthquake. Then, there was a rally into June, followed by a sharp decline to low in July, with the bottom just slightly higher than the low of May. From this low, there was an advance to September, when another top was made, but lower than the top in January; then followed a decline into December and a panic followed in 1907.

1926 — The next important 10-year cycle to consider is 1926, when the great Coolidge bull campaign was under way. From the low in December 1925, stocks rallied to February 1926 then had a sharp decline to March, some stocks breaking as much as 100 points. From this bottom, there was a sharp advance to new high levels, reaching top in August; then another sharp decline to bottom in October, from which a rally followed to December, but stocks did not get back to the high until August that year.

Now, when I get ready to make up my forecast for 1936, I will consider all of these cycles. I will go back and also check the 7-year, the 14-year and 15-year cycles, which is half of the 30-year cycle. However, at this writing, with my knowledge and experience of the future cycles, I expect the 1896 cycle to repeat in 1936.

1936 is likely to be a very uncertain election year just as it was in 1896, when the Bryan silver scare caused a panicky decline into August. There is a possibility of a three-cornered fight, with two Democratic presidential candidates and one Republican. There certainly is going to be a time during 1936 when the investors are going to get scared and speculators are going to get scared and sell stocks, causing sharp declines.

My opinion, at this writing, is that the first decline will start in the month of January and wind up with a sharp decline. In February, the market may drift along in a narrow, trading range with some rallies, but there will be another decline in the month of March, just as there was in 1926. I am confident that there will be another break in the months of May and June, especially in the latter half of May, as this will be running out four years from the 1932 low and 6 years from April 1930 high, all of which are indications of important changes in trend.

We know that presidential nominations will take place in July; therefore, this is a month to watch for uncertainties and declines, unless sharp declines have come before that time. The ending of the cycle from 1896 in August is quite important and regardless of how high or low stocks are, there are likely to be some sharp declines before the end of August. Again, in the last half of September, uncertain conditions and possibility of sharp declines are indicated. This may mark the last low and an election rally may start if there are indications of a change in administration by the election of a Republican president, which, I believe, at this writing, will happen.

September, October, and November are all important because these months are 7 years from the top in September 1929 and 7 years from the panicky decline in October and November 1929. I would expect a rally to take place after the election in November, which would last anyway until the early part of December. If conditions show signs of improvement and if people are satisfied with the man elected, and then the advance will probably continue into December, with high prices around the end of the year.

This is merely a general outline I am giving without completing all of my calculations and making up the Annual Forecast in detail.

INDIVIDUAL STOCKS:

I have told you before that you should not depend upon the averages to forecast the trend of individual stocks. These averages give you the general trend, and while many stocks will follow this average trend, you should figure out each stock individually and let its position on geometrical angles and time periods determine the different months in the year when the stock is likely to make many tops and bottoms.

Take any individual stock and make up a chart like the Master Forecasting Chart, carrying it across 10 years or 20 years, and see how its tops and bottoms come out. I have made up a chart of the 10-year cycles on U. S. Steel and also a chart of the 20-year cycles, and I am always glad to furnish these charts to students of my course on forecasting so that they may study the individual stocks and be convinced that the theory will work on an individual stock even better than it will work on the averages.

No man can study the Master 20-year Forecasting Chart and the cycles without being convinced that the time cycles do repeat at regular intervals and that it is possible to forecast future market movements. By studying resistance levels, geometrical angles, and volume of sales in connection with the cycles, you can determine when the trend is changing at the end of campaigns.

FAST MOVES AND CULMINATIONS AT IMPORTANT TIME PERIODS

It is important to go over the monthly chart of Industrial or Railroad averages or any individual stock and look up the months when fast advances and fast declines have occurred and figure the number of months from any important top and bottom.

Watch how bottoms and tops come out on the important geometrical angles or proportionate parts of the circle of 360 degrees, which are:

11 ¼	56 ¼	*90	123 ¾	168 ¾	213 ¾	247 ½	292 ½	326 ¼
22 ½	*60	101 ¼	*135	*180	*225	258 ¾	*300	337 ½
33 ¾	67 ½	112 ½	146 ¼	191 ¼	236 ¼	*270	303 ¾	348 ¾
*45	78 ¾	*120	157 ½	202 ½	*240	281 ¼	*315	*360

(*very important)

These angles measure the time periods. Always watch what happens around 45, 60, 90, 120, 135, 180, 225, 240, 270, 300, 315, and 360 months from any important top or bottom, as all of these angles are very strong and important, just the same as the 45-degree angle, and indicate strong culmination points.

REVIEW OF DOW-JONES INDUSTRIAL STOCKS FROM 1896:

Go back to the extreme low of August 1896 —

1897 — A secondary low was recorded in April 1897. We find that there was a fast advance in the 13th to 13th months from August 1896 low.

1898 — A fast advance occurred in the 16th and 24th months from the bottoms of 1897 and 1896, and a fast decline in the 17th and 25th months.

1899 — A bull year, fast advance occurred in the 29th to 32nd months from 1896 and in the 21st to 24th month from the 1897 bottom. Fast declines occurred in the 40th and 32nd months from these bottoms.

1900 — Fast advance 42nd to 44th month from 1897 and 50th to 52nd months from 1896 bottom.

1901 — A fast decline on the 49th month from 1897 and 57th month from 1896 low. Top reached in June.

1903 — A bear year. In the 22nd to 28th month from 1901 top, a fast decline — also 72nd to 78th months from 1897 bottom and 80th to 86th months from 1896 bottom. Bottom reached in October and November 1903.

1904 — Fast advance, 12 to 14 months from 1903 bottom.

1905 — Fast moves up in the 16th to 18th months; fast decline in the 19th month and a fast advance in the 25th to 27th months from 1903 bottom.

1906 — Top of campaign reached in January. Fast decline in the 30th month from 1903 bottom.

1907 — Fast decline in the 14th month from 1906 top and in the 19th to 22nd months. Extreme low reached in November 1907, in the 135th month from 1896 bottom, 127 months from 1897 low, and 22 months from 1906 top.

1909 — Top of campaign reached in October 45 months from 1906 top and 23 months from 1907 bottom, 158 months from 1896.

1914 — July, a fast decline in the 57th month from 1909 top, 21 months from 1912 top. Extreme low of campaign in December, 107 months from 1906 top, 26 months from 1912 top, 220 months from 1896 low, 84 months or 7 years from 1907 bottom, and 134 months from 1903 bottom.

1915 — This was a war year. March and April — Fast advance on the 3rd and 4th month from the 1914 bottom. May —A sharp, severe decline, 90 months from November 1907 bottom and 225 months from 1896 bottom. Note these fast moves on a 90-degree angle and 225-degree angle, which is equal to a 45-degree angle, or 180 plus 45.

1916 — April — A sharp decline, 16 months from the 1914 bottom, 123 months from 1906 top, and 236 months from 1896 low. September—Fast advance, 21 months from 1914 low, and 240 months from 1896 low, the end of the 20-year cycle, indicating an important change in trend. November —Top of a fast advance. Dow-Jones Industrial Averages at the highest price in history up to that time. This was 23 months from 1914 bottom and 243 months from 1896 bottom. December—A sharp decline, 24 months from 1914 bottom.

1917—August to December — A fast decline, 9 to 13 months from November 1916 top, 32 to 36 months from the 1914 bottom, 117 to 121 months from the 1907 bottom, and 252 to 256 months from 1896 low.

1919 — A fast advance started in February and lasted until July. This was 27 to 32 months from the 1916 top, and 50 to 55 months from 1914 low. February 1919 was 135 months from the 1907 low and 270 months from 1896 bottom. The 135th and 270th months, being 3/8 and $^{3}/_{4}$ of the circle were very important for changes in trend and starting of moves. October and early November — Final top, 36 months from 1916 top. November — A panicky decline, 23 months from 1917 low, 59 months from 1914 bottom (end of a 5-year cycle), and 279 months from 1896 bottom.

1920 —November and December —A fast decline, 12 to 13 months from 1919 top, 35 to 36 months from 1917 low, 72 months from 1914 bottom, 157 months from 1907 bottom, and 291 to 292 months from 1896 bottom.

1921 — August — Low of bear campaign, 21 months from 1919 top, 80 months from 1914 bottom, 165 months from 1907 bottom, and 300 months from 1896 bottom.

1924 — May — The last low was made, from which a fast advance started one of the greatest bull campaigns in history, ending in 1929. This was 54 months from the 1919 top, 33 months from 1921 low, 113 months from 1914 low, and 333 months from 1896 low.

1926 — March — A big decline, with some stocks declining 100 points. This was 23 months from May 1924 low, 29 months from 1923 low, 55 months from 1921 low, 135 months from 1914 low, and 355 months from 1896 low. August — Stocks reached the highest price up to that time, the Dow-Jones Industrial Averages selling at 166. This was 27 months from May 1924 low, 34 months from October 1923 low, 60 months from 1921 bottom, 225 months from 1907, and 360 months or 30 years from 1896 low. Then a 20-point decline followed to October, which was 2 months in a new 30-year cycle from the bottom of 1896.

1928 and 1929 were years of some of the fastest moves in history.

1929 — May to September — One of the fastest moves, advancing nearly 100 points on averages. Final high in September. This was:

118 months from 1919 top.　　　　　　97 months from August 1921 low.

240 months from 1909 top.　　　　　　177 months from 1914 low.

42 months from March 1926 low.　　　262 months from 1907 low.

64 months from May 1924 bottom.　　37 months in the second cycle of 30 years from 1856 low.

71months from October 1923 low.

Note the strong time angles on the monthly chart running out in October and November, 1929, which are 32, 40, 45, 67 $^1/_4$, 75, 120, 180.

1930 — April — Last top before another big decline. This was 49 months from March 1926 low, 71 months from 1924 low, and 78 months from 1923 low. May — A sharp decline. This was 270 months from 1907 low and 45 months in the second cycle from 1896 low. Then there were fast declines to 1931.

1931 — September — A decline of 46 points on the Dow-Jones Averages. This was 24 months from the 1929 top, 95 and 86 months from 1923 and 1924 lows, 121 months or the beginning of a new 10-year cycle from the 1921 low, 201 months from 1914 low, and 61 months in the new cycle from 1896.

1932 — July 8th — Extreme low of the bear campaign was reached. This was 71 months from in the new cycle from 1896 low. 131 months from 1921 low, 105 and 96 months from 1923 and 1924 lows, 27 months from April 1930 top, and 34 months from the 1929 top. August and September — A sharp, fast advance in stocks. This was 35 and 36 months from 1929 top, 28 and 29 months from April 1930 top, 72 and 73 months in the new cycle from 1896 low, and 132 to 133 months from 1921 low.

1933 - April to July — A fast advance. This was 43 to 45 months from the 1929 top. Always watch for culminations around the 45th month and multiples of 45. It was also 36 to 39 months from the 1930 top, 9 to 12 months from 1932 low, and 80 to 83 months in the new cycle from 1896, or running out a 7-year cycle in the new 30-year cycle. October 1933 — Low of reaction, 42 months from April 1930 top, 49 months from 1929 top, and 15 months from 1932 low.

1934 — February — Top. This was 46 months from 1930 high, 53 months from 1929 high, 12 months from 1933 low, 19 months from 1932 low, and most important of all, 90 months into the new 30-year cycle from August 1926. From this top, a sharp decline followed. July — This marked the last low before a big bull campaign started. This was 58 months from 1929 top, 51 months from the 1930 top, 24 months from 1932 low, and 95 months or ending of the 8th year in the new cycle from 1896. Going into the 9th year of this cycle, the market indicated a big bull campaign to follow in 1935 as explained before.

Go over individual stocks and work out their cycles in the same way. Look up the months when extreme highs and lows have been made and note the months from each bottom and top when fast advances and fast declines have taken place. By keeping up the time periods from important tops and bottoms, you will know when important time periods are running out and when a change in trend is likely to take place. Also watch the seasonal changes in trend around March to April, September to October, and November to December.

All of this will help you to pick the stocks that are going to have the greatest advances and the ones that are going to have the greatest declines. The more you work and study, the more you will learn and the greater profits you will make.

NEW YORK STOCK EXCHANGE PERMANENT CHART

This master chart is a square of 20, or 20 up and 20 over, making a total of 400, which can be used to measure days, weeks, months or years, and to determine when tops and bottoms will be made against strong angles as indicated on this permanent chart. This chart works out the 20-year cycles remarkably well because it is the square of 20. For example:

The New York Stock Exchange was incorporated on May 17, 1792. Therefore, we begin at "0" on May 17, 1793. 1793 ends on "1," when the Stock Exchange was one year old. 1812 will come out on 20, 1832 on 40, 1852 on 60, 1872 on 80, 1892 on 100, 1912 on 120, and 1932 on 140. Note that 140, or 7 times 20, in 1932 is equal to a 90-degree angle and is at the top of the 7th zone or the 7th space over, which indicated that 1932 was the ending of a bear campaign and great cycle and the starting of a bull market. We would watch for a culmination around May to July 1932 and the cycle ended May 17, 1792.

You will notice that the numbers which divide the square into equal parts, run across 10, 30, 50, 70, 90, 110, etc., and that the year 1802 comes out on 10, the year 1822 on 30, the year 1842 on 50, the year 1862 on 70. Note that the year 1861, when the Civil War broke out, was on the number 69, which is on a 45-degree angle. Then note that 1882 ended in May on the 90-degree angle and at the $\frac{1}{2}$ point, 180-degree angle, running horizontally across.

Again in 1902 it was at 110, the $\frac{1}{2}$ point, and in 1903 and 1904 hit the 45 degree angle. Note that the years 1920 and 1921 hit the 45-degree angle on No. 129, and 1922-the first year of the bull market- was at 130 at the $\frac{1}{2}$ point.

Note that 1929 was on the 137th number, or 137 month, and hit an angle of 45-degrees, and that the year 1930 was at the $\frac{1}{2}$ point on the 4th square, a strong resistance point, which indicated a sharp, severe decline.

1933 was on 141 or the beginning of the 7th zone and at the center or halfway point of the 2nd quarter of the square of 20, indicating activity and fast advances and fast declines.

The years 1934 and 1935, ending in May, were on 142 and 143, and 1936 comes out on the 45-degree angle at the grand-center in the 8th Zone and at the half-way point of the 2nd square, going to $\frac{1}{2}$ of the total square, which indicated great activity.

You can also use this chart from October 12, 1492, when Columbus discovered America. 1892 was end of 400 years or square of 20. 1932 was 40 years in the new square of 20.

You can use this square of 20 for time periods on individual stocks and for price resistance levels.

If you will study the weeks and months, as well as the years, and apply them to these important points and angles, you will see how they have determined the important tops and bottoms in the past campaigns.

1	2	3	4	5	6	7	8	9	10	11	12	13	14	15	16	17	18	19	20
20	40	60	80	100	120	140	160	180	200	220	240	260	280	300	320	340	360	380	400
19	39	59	79	99	119	139	159	179	199	219	239	259	279	299	319	339	359	379	399
18	38	58	78	98	118	138	158	178	198	218	238	258	278	298	318	338	358	378	398
17	37	57	77	97	117	137	157	177	197	217	237	257	277	297	317	337	357	377	397
16	36	56	76	96	116	136	156	176	196	216	236	256	276	296	316	336	356	376	396
15	35	55	75	95	115	135	155	175	195	215	235	255	275	295	315	335	355	375	395
14	34	54	74	94	114	134	154	174	194	214	234	254	274	294	314	334	354	374	394
13	33	53	73	93	113	133	153	173	193	213	233	253	273	293	313	333	353	373	393
12	32	52	72	92	112	132	152	172	192	212	232	252	272	292	312	332	352	372	392
11	31	51	71	91	111	131	151	171	191	211	231	251	271	291	311	331	351	371	391
10	30	50	70	90	110	130	150	170	190	210	230	250	270	290	310	330	350	370	390
9	29	49	69	89	109	129	149	169	189	209	229	249	269	289	309	329	349	369	389
8	28	48	68	88	108	128	148	168	188	208	228	248	268	288	308	328	348	368	388
7	27	47	67	87	107	127	147	167	187	207	227	247	267	287	307	327	347	367	387
6	26	46	66	86	106	126	146	166	186	206	226	246	266	286	306	326	346	366	386
5	25	45	65	85	105	125	145	165	185	205	225	245	265	285	305	325	345	365	385
4	24	44	64	84	104	124	144	164	184	204	224	244	264	284	304	324	344	364	384
3	23	43	63	83	103	123	143	163	183	203	223	243	263	283	303	323	343	363	383
2	22	42	62	82	102	122	142	162	182	202	222	242	262	282	302	322	342	362	382
1	21	41	61	81	101	121	141	161	181	201	221	241	261	281	301	321	341	361	381

FORECASTING WITH GANN'S CYCLES

The reason for using planets as a measurement for timing potential changes in market trends is that they are natural occurring units of time. This is not astrology! Many people will laugh at this concept and say that it is a ridiculous theory. However, these same people have no problem in accepting standard measurements of time such as days, weeks, months, and years. This is why Gann did not talk or write about astrology or planetary cycles even though it was his main forecasting tool. If we examine the typical measurements of time, we will see that they are all based upon planetary movement. It just so happens to be the planet we live on.

102

A day or 24 hours is simply the amount of time it takes the Earth to rotate or spin 360 degrees on its axis. A week is based on the lunar quarters; a month is the amount of time it takes the Earth to move through a zodiac sign, or 30 degrees, which also completes a 360-degree revolution of the moon around the Earth. Finally a year is the amount of time it takes the Earth to complete an orbit around the sun. Our entire concept of time is based upon planetary movement. So why should it seem "weird" to use other naturally occurring time measurements! Is it so far fetched to believe that cycles are actually caused by planetary orbits? Even if the planets do not cause cycles, the orbits still represent a natural unit of time that may have some universal harmony that we are unaware of. Whatever your personal beliefs are, the fact is that Gann used planetary orbits to time market movements. In the information above, you have a reprint of *Gann's Forecasting Course.* Gann produced two different forecasting courses; this one is rarer than the other one titled *Forecasting by Time Cycles,* which is typically found in Gann's Master Courses for either Stocks or Commodities. In the course provided, Gann gives very specific rules for figuring out when the next top or bottom should be expected. It is important to know that he is not talking about the typical calendar units of time. Since Gann's rules for using these time cycles is clearly laid out in the course, I am not going to discuss how to use these cycles. What I am going to do is define in my own opinion about the actual planetary measurements that Gann was really pointing the student towards increasing accuracy. All of these cycles are heliocentric orbits, not geocentric. Because of the elliptical nature of planetary orbits and changing speeds due to perihelion and aphelion, these important time periods do not typically equal the fixed units of time that Gann wrote about in the course. This is why you must learn how to use these planetary clocks to master Gann timing techniques!

GREAT CYCLE - 60 YEARS: This cycle is the third conjunction of the two largest planets in the solar system, Jupiter and Saturn. A conjunction between the two massive planets takes place every twenty years and is also Gann's 20-year cycle. (The appendix has a chart from Man in Cosmos that illustrates the importance of conjunctions of Jupiter and Saturn on repeating patterns in the Dow Jones Industrial Average. Note: This chart centers the calendars of the price charts with geocentric Jupiter conjunct the Sun. This time period is also two complete Saturn orbits.

50-YEAR CYCLE: This cycle of 49 to 50 years (a period of "jubilee") is related to 210 degrees of Uranus which moves 30 degrees in 7 years. Other potential cycles are $3\frac{1}{2}$ conjunctions of Jupiter & Uranus (every 14 years) or the geometric mean of Saturn and Uranus, which is also equal to this time period. Gann also describes this as being 225° to 240° in the second circle of 360°, the period of being 586 months.

30-YEAR CYCLE: This is the orbit of Saturn.

20-YEAR CYCLE: This is the conjunction of Jupiter and Saturn. This period is also 2/3 of Saturn's orbit.

15-YEAR CYCLE: This is $\frac{1}{2}$ Saturn's orbit also conjunctions of Jupiter and Uranus (14 yrs.) should be watched.

10-YEAR CYCLE: 1/2 cycle of Jupiter/Saturn conjunctions. Also = 120 degrees of Saturn's orbit.

7-YEAR CYCLE: 30 degrees of Uranus and also $^1/_2$ cycle of conjunctions with Jupiter and Uranus.

5-YEAR CYCLE: This is the secret of the hexagon chart. Each 5-year period is based on 60 degrees of Saturn completing all six sides of a cube or hexagon. It is also $^1/_4$ of the Jupiter/Saturn conjunction cycle.

MINOR CYCLES 3 AND 6 YEARS: This is 90 and 180 degrees ofJupiter, respectively.

2-YEAR CYCLE: One Mars/Saturn conjunction, one Mars orbit 1.88 years.

7-MONTH CYCLE: One Venus orbit is 225 days.

3-MONTH CYCLE: One Mercury orbit (such as 10/18/99 — 1/14/00 Dow Jones Mercury moved 360 degrees). This is also 90 degrees of the Earth's orbit.

1-MONTH CYCLE: One lunar orbit around the Earth, 30 degrees of the Earth and 1-degree of Saturn

1-WEEK CYCLE: 90 degrees of the Moon's orbit.

Gann goes on to say: "Stocks move in 10-year cycles, which are worked out in 5-year cycles —a 5-year cycle up and a 5-year cycle down. Begin with extreme tops and extreme bottoms to figure all cycles, either major or minor." From here on, he explains his rules for adding specific time periods to tops and bottoms. This is where you need to use the planetary periods above to calculate the future turning point.

You should pay close attention to Gann's instructions because he is specific about whether the future date is supposed to be a top or bottom. Also, note that you may get conflicting cycles, i.e., one cycle may be forecasting a top and another bottom. This is why Gann said: *"Time is more important than price."* Because when the time element comes in you will be able to visually determine if the market is topping or bottoming from its price action and trend. Cycles are additive, so you must weigh the total number of topping cycles against the total number of bottoming cycles to be completely accurate in forecasting a top or bottom. This is really not important as far as the timing is concerned.

IMPORTANT FORMULAS AND TECHNIQUES
FOR WORKING WITH PLANETARY CYCLES

First, it is a necessity that you learn and memorize the zodiac signs, zodiac symbols and degrees of the circle, and the planetary symbols. Next, you should memorize the time required for a planet to complete an orbit around the sun. You should also memorize the time period required for conjunctions of the outer planets with each other. To illustrate the importance of this lesson, I will provide a few simple examples.

At the stock market low, August 12, 1982, Saturn was at 23 degrees Libra. Adding 180 degrees to this position gives 23 degrees Aries, which was hit April 22, 1998 where the vast majority of Stock Market Indexes topped before correcting over 20%. In addition, Jupiter was at 13 degrees 45' Scorpio on August 12, 1982. Adding 540-degrees (360 + 180) yields 13-degrees 45' of Taurus, which will be hit on March 23, 2000. This is the exact top of the 18-year bull market. Also notice that heliocentric Mars and Saturn are conjunct at this time, completing a two-year cycle. In addition, Mars was at 18-Taurus completing his hexagon cycle.

The Asian Currency Panic occurred on October 22, 1997 with Mars at 18 degrees Capricorn. Adding 180 degrees (60+60+60) to Mars gives 18 degrees Cancer, which was the Russian Currency Panic August 31, 1998! This is Gann's 1-year cycle that often shows a change in trend in the 10th or 11th month.

Now we will look at an example with the relationship of two planets. The first thing you must remember is that a pair of planets can never be more than 180 degrees apart. Second, you must observe if the planets are getting closer together (applying) or getting farther apart (separating). I will give an example of each; both will be updates from Four-Dimensional Stock Market Structures & Cycles by Bradley Cowan.

At the market top January 29th, 1994 Saturn is @ 2 06' Pisces and Uranus is @ 22 28' Capricorn. The angle between the two is 332.10 — 292.47 = 39.63 degrees or 39-degrees 37' minutes of separation. The planets are getting farther apart moving towards 180 or opposition. Adding 30 degrees to this amount of separation gives 69 37' which was the top of the S&P 500 index October 7th, 1997 just prior to the Asian Currency Panic. Adding another 30 degrees gives 99 37', which will hit on January 28, 2001. The next significant turn will come at $7^1/_2$ ($^1/_2$ of 15) + 99 37' = 104 37' separation or 7/25/2001. You can keep adding harmonics of 30-\ degrees (15, $7^1/_2$, 3%) or harmonics of 45-degrees (22 $^1/_2$, 11 $^1/_4$, 5 5/8ths) to calculate future cycle turning dates. Once the number gets to 180, you have to subtract because the planets are no longer separating; they are applying. For example 90 degrees + 89 37' = 179 37'. If we want to calculate the next 30-degree cycle date we would do the following: 180 —179 37' = 23' minutes. This is the amount of movement left before the planets start applying. 30 degrees minus 23' minutes = 29 37'. This is the amount we have to subtract from 180 (maximum separation) to calculate the next angle relationship. 180 — 29 37' = 150 23'. This example shows the math required for both separating and applying planet scenarios. Remember that there are only 60' minutes in a degree.

This process is very similar to calculating time on a non-digital clock i.e., the old fashion second, minute and hour hands. If you try to visualize the fast moving planets as the second hand and the slower planets as the minute and hour hands, you will have a pretty good understanding of what is physically happening between the planets. The next section is an article I wrote several years ago that should further help with your understanding of planetary cycles and market timing.

W.D. GANN'S SECRETS TO FORECASTING

Today there are so many "Gann Experts" that are cashing in on the mysterious methods of W.D. Gann. There are tons of books, several courses, computer programs, trading systems or methods, etc., that lure in the trading public, especially commodities traders, incouraging them to spend their hard earned money in the hopes of discovering the true "Holy Grail" techniques of Mr. Gann. I have personally spent countless sums of money over 15 years or more of personal study, in the pursuit of Gann's methods. I really have no idea what the total final bill amounts to, but I can easily recall how much of it was a value and how much of it was complete junk. I won't slam anyone or point the finger at authors of material that is of little value. What I am willing to do is share my interpretation of Gann's work with the public in the form of a book. I never planned to write a book. But after all of the letters I received from the two previous articles that I wrote for Traders World requesting that I offer more information, I decided to put some information together that would be truly valuable to the "Holy Grail" seeker. To prove my sincerity, and the value of this work, I will provide you with extremely valuable information in this article on forecasting.

The main attraction with Gann is forecasting! The idea that the future can be predetermined is a fascinating concept. To truly understand Gann's methods, you must understand his concept of time. Gann said that *"Time Proves All Things"* and that *"Every movement in the market is the result of a natural law and a cause which exists long before the effect takes place and can be determined years in advance. The future is but a repetition of the past, as the Bible plainly states."*

"The thing that hath been, it is that which shall be; and that which is done is that which shall be done, and there is no new things under the sun." Eccl 1:9.

Gann's Forecasting Course deals with the use of "Major Time Cycles." Gann believed that everything moves in cycles (circles) as a result of the natural law of action and reaction. He has said, " By a study of the past, I have discovered what cycles repeat in the future." The problem that most students of Gann run into is that they get caught up in trying to predict price. Gann always said that between the two, time was much more important than price. Knowing when to trade is the key to Gann. I will now provide you with a short term and a long-term example based upon the instructions: "Rules for Future Cycles" by W.D. Gann:

Rule #4- Adding 10 years to any top, will give you the top of the next 10-year cycle, repeating the same kind of year and about the same average fluctuations

To really understand what Gann meant by this rule, you have to understand his concept of time or what he means by cause and effect. Gann was a master mathematician, mason, and also an astrologer! In fact, most of Gann's recommended reading list deals with astrology. Now, let's look at what 10 years means to Gann.

Primarily, 10 years is equal to $\frac{1}{2}$ of a Jupiter and Saturn cycle. Secondly, 10 years is equal to 120-degrees of Saturn or two sides of the cube on his hexagon chart.

At the market top of August 25th, 1987, you will find that heliocentric Saturn has a longitude of 20 degrees 12' minutes Sagittarius and that heliocentric Jupiter's longitude is 19 degrees 44' minutes of Aries. The angle between these two planets is 119 degrees 32' minutes and they are separating. If we add 120 degrees to Saturn (two five-year sides of the hexagon), we get 20 degrees 12' minutes Aries, which occurred on January 13,, 1998. If we use the angle between Jupiter and Saturn, then we must find out when these two planets are 60 degrees apart before the conjunction i.e., 180 degrees minus 119 32' = 60 degrees 28' minutes. This occurred around July 16, 1997. If you followed this cycle as it was presented in Bradley Cowan's book *Four-Dimensional Stock Market Structures and Cycles* you would have been looking for an angle of 58 degrees, which happened on August 6' 1997. Brad's cycle work is an excellent resource for understanding Gann's Forecasting Course. This example illustrates the top-to-top relationship that Gann presented as rule #4 in his forecasting course. Both dates were exact hits!

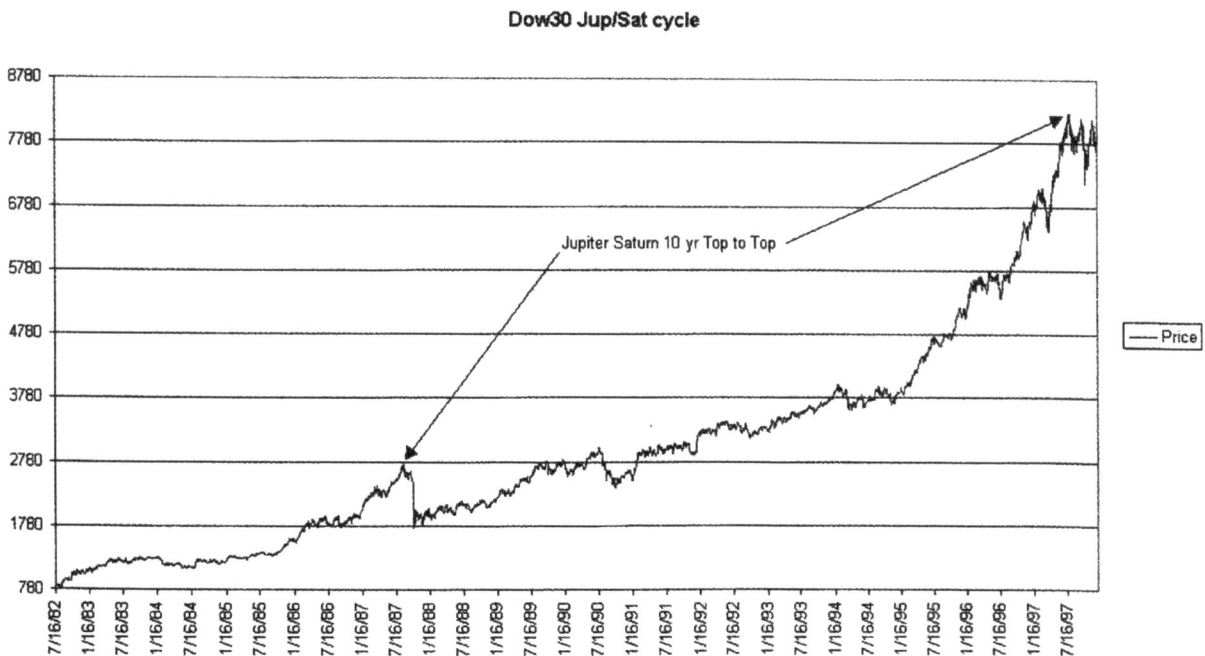

Dow30 Jup/Sat cycle

FORECASTING MONTHLY MOVES

A short-term example or forecasting rule is to add three months to an important bottom, then add 4, making 7, to get minor bottoms and reaction points.

Again, we have to decode Gann's definition of time. Three months is 360 degrees of Mercury. Adding 4, making 7 months is 360 degrees of Venus. If you examine the recent low of October 18, 1999, you will find heliocentric Mercury at 21 degrees Capricorn and Venus is at 1-degree 26' minutes of Gemini. Looking in an ephemeris, we will find Mercury again at 21 Capricorn on January 14, 2000 (the exact top of the Dow Jones before a 15% + correction). In addition, Venus returns to 1-26' Gemini on Tuesday, May 30, 2000 for another potential turning point. At the time of this writing, March 15, 2000, the Venus forecast is yet to be fulfilled so watch the market around this day! I hope that this information proves to be valuable in your quest for knowledge, as this will be my last article. One final parting piece of wisdom is that when you think of cycles, think of throwing a stone into a pond of water. Each turning point in the market starts a ripple effect that moves forward in time and affects the future. The more significant the turning point, the larger the ripples. Just like throwing in a small pebble and a large rock. It is obvious that the large rock will overpower the ripples of the small pebble, making its effects non-distinguishable at times. In these instances, record the position of the minor cycles at the time of the "major" turn and project these values into the future, along with the old values, to see if the minor cycle has remained unchanged or has synchronized itself with the larger cycles, "wheels within wheels." This is why Gann said to work all cycles from major tops and bottoms in the market.

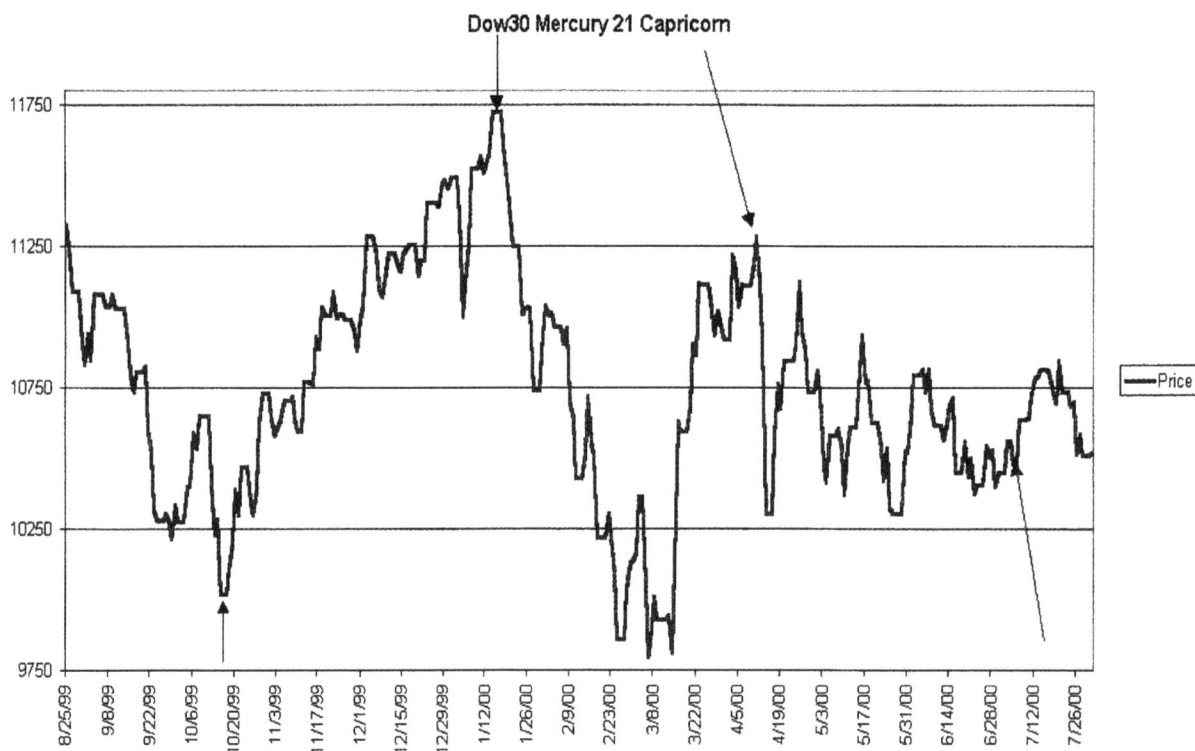

Dow30 Mercury 21 Capricorn

Also pay attention to April 12, 2000 and July 7, 2000, when Mercury hits 21-Capricorn for the 3' and 4th times, etc. The chart above has been updated so you can now see what happened.

Bonus Tip: Watch the market for trend changes when Helio-Venus transits: 2-Aries, 17-Taurus, 2-Cancer, 17-Leo, 2-Libra, 17-Scorpio, 2-Capricorn and 17-Aquarius.

THE MASTER PRICE AND TIME CHART
SQUARES 1 TO 33 INCLUSIVE
PRICE AND TIME 1 TO 1089

The Square of Nine is very important because nine digits are used in measuring everything. We cannot go beyond 9 without starting to repeat and using the 0. If we divide 360° by 9, we get 40, which measures 40°, 40-months, 40-days, 40-weeks and shows why bottoms and tops often come out on these angles measured by one-ninth of the circle. This is why the children of Israel were 40-years in the wilderness.

If we divide our 20-year period or 240 months by 9, we get 26-2/3 months, making an important angle of 26-2/3°, months, days or weeks. Nine times nine equals 81, which completes the first Square of Nine. Note the angles and how they run from the main center (81 is on a 45°angle from the center with all the other "odd" number squares). The second Square of Nine is completed at 162. Note how this is in opposition to the main center (163 is in opposition to the center, 162 is 1 digit off). The third Square of Nine is completed at 243, which would equal 243 months or 3 months over our 20-year period and accounts for the time, which often elapses before the change in cycle, sometimes running over 3 months or more. The fourth Square of Nine ends at 324. Note the angles of 45° cross at 325, indicating a change in cycles. To complete the 360° requires four Squares ofNine and 36 over. Note that 361 equals a square of 19 times 19, thus proving the great value of the Square of Nine in working out the important angles and proving discrepancies.

Beginning with "1" at the center, note how 7, 21, 43, 73, 111, 157, 211, 273, and 343 all fall on a 45° angle. Going the other way, note that 3, 13, 31, 57, 91, 133, 183, 241, and 307 also fall on an angle of 45°. Remember there are always four ways you can travel from a center following an angle of 45°, or an angle of 180° (3-dimensional objects), or an angle of 90°, which all equal about the same when measured on a flat surface. Note that 8, 23, 46, 77, 116, 163, 218, 281, and 352 are all on an angle from the main center; also note that 4, 15, 34, 61, 96, 139, 190, 249, and 316 are on an angle from the main center, all of these being great resistance points and measuring out important time factors and angles.

We have astronomical and mathematical proof of the whys and wherefores and the cause of the workings of geometrical angles. When you have made progress, proved yourself worthy, I will give you the Master Number and also the Master Word.

This chart starts with the square of 1 in the center, and moves clockwise around with the odd squares coming out on the 45-degree angle. These numbers are 1, 9, 25, 49, 81, etc. The even squares run in the opposite direction on a 45-degree angle, beginning with the square of 2, which is 4 and continuing on this angle (16, 36, 64, 100, etc.) This produces a variable in time and price of 2. That is 2 points in price, 2 days, 2 weeks, or two months in time. This chart proves why prices move so much faster at higher levels, and measures exact resistance levels in the squares.

Example: May soybeans extreme low 44 cents. This is in the square of 7, from 43 to 49 is 90 degrees. When the price was at 436-³/₄, from 421 to 441 covered 90 degrees. Therefore, to swing between these angles required 20 points, while at the lower levels it was only 5 points. It is the same with the time periods. At the present time, May soybeans is in the 253rd month from December 28, 1932 and you will note that from 241 to 257 is 90 degrees, or 16 points, in price or 16 periods in months, weeks and days. You will note that the 253rd months is on the angle of 22½ degrees or 112½ degrees from the starting point and the opposite point of this angle is January 13, making January 13 to 15 important for a change in trend. The time periods starting at the left and in the east beginning at March 21 are the seasonal time periods, and to get the same position on time, you would start soybeans from December 28, which is just a little past the seasonal date of December 21, and January 15, which is just 2 days from the 22½ degree angle, July 27, extreme low on May soybeans is just beyond July 14, where the 22 ½ degree angle comes out.

All of the important highs on May soybeans are marked with a green circle. The important lows are marked with a red circle. You will note that 44 cents was just one point away from the 45-degree angle (43) and that 436³/₄ the extreme high was on this 45-degree angle. Also on a green angle of 22'2 degrees. The extreme low on May beans, 67, on July 27, 1939, was on a green angle of 22½ degrees, and this angle runs to the date of August 31st, and February 28th. Also 202½ degrees was on the same line with 67, and 405 was just 1-cent away from the angle of 221½ degrees or 406. From the important highs and lows, I have drawn 45-degree angles and 90-degree angles so you can see the important resistance levels.

Example: The recent high of 311¹/₄ on May soybeans made on December 2, 1953, was on the 90-degree angle or straight up from 436³/₄, and also on the angle of 22½ degrees, which runs from 44 and 277, and you will note that 310 is on a 45-degree angle from 240 the low in August 1953, making this a strong resistance and selling level. The time period of 253 months is in red figures and the price of 305 is on the 90-degree angle from 233 low and 240 low. A price of 305 is below the 45-degree angle from 44, 344, and 240. It is on the angle of zero degrees from 240. When the price sells at 303 it will be below the 45-degree angle from 240. A complete cycle or round trip is most important to watch for a change in trend. From 240 to 305, was a complete square, cycle or round trip, but to reach on the 90-degree angle, the price had to make 308. The natural resistance level from the 45-degree angle at 307, to the 90-degree angle at 316, or on half was 311, the natural resistance and selling level.

When May soybeans declined on December 17th to 296, they were on the 45-degree angle from 44 cents, because the time period was 252 months and we add 44, which gives 296, making this a temporary support and buying level. Also, it was 1-cent above the angle of no degrees or 180-degrees east of 44-cent extreme low.

You should always consider how many degrees the price has moved from an extreme high to an extreme low. From a high of 311 to 298, is 67½ degrees and is about 11¹/₄ degrees, which would make 78³/₄ degrees or 7/8th of 90.

When the price had advanced from 240 to 305, it had moved 360 degrees or a complete circle. Therefore, at 311, it had moved $33^3/_4$ degrees more than the circle of 360 degrees. For the price to decline to the next natural resistance level from 340 low would be 289, which would be on a 90-degree angle and on the square of 17 and on the 45-degree angle in the natural squares. To move to 90 degrees from 311 would be 285. The next important resistance level would be 277-276, which would be 180 degrees from 311 and on the same angle of $22^1/_2$ degrees.

Bring up all time periods from monthly highs and lows and weekly highs and lows and see how they stand in the squares in relation to the price.

Example: For the week ending January 9, 1954, May soybeans will be in the 20th week from the August 20th, low. Note that the beginning with square of **1** at 20 on the angle of $22^1/_2$ degrees and should the price drop below 303 it will be below this time angle, and should it decline to 297 it will be on no degrees or 180 degrees from 20, in the time period.

February 15, 1920, high 405. November 15, 1953 was 405 months, therefore, December 15 was 406 months and January 15th will be 407 months and 303 is no degrees or 180 degrees from this angle.

January 15, 1954 to February 15th is the 73rd month and this is on a 45-degree angle naturally, making February important for a change in trend.

1959, July 27, May beans low, 67. To January 27, 1954, will be 174 months. Note that 174 is opposite the price of 296, and that the 176th month will be March 27th, and this will be in the balance between the two red lines and that the seasonal time period is marked March 21, making this important to watch for a change in trend. Suppose the price is at 288, this will be on a 90-degree angle of the time period of 176. And 176 of course, is opposite the 69 low and 178 months, which will be May 27th and will be opposite 180 degrees from 67, the extreme low on May beans.

If you will take the time to study and practice with this master chart, using all of the time periods and price levels, you will soon find that it is easy to determine a change in trend from this chart alone.

The spiral chart represents the correct position, time and space of anything that begins at zero and begins to move round and round. It shows just exactly how the numbers increase as the spiral moves round and round and why stocks move faster as they grow older, or swing so much more rapidly as the price reaches higher levels. At the center, beginning point or zero, it requires 45° to represent one point. When the stock has traveled seven times around the center, it then requires seven points to strike a 45° angle. When it has traveled around the spiral twelve times, it will then require a space of ten points before striking a 45° angle. It would also mean that the stock could move in one direction ten months without striking anything to cause any great reaction. On this chart, we have only shown the 45, 60, 90, 120, 135, 180, 225, 240, 270, 300, 315, and 360-degree angles. This shows the division of the circle by 2, 4, and 8, and also shows the one-third point and the two-thirds point; being the vital and most important angles, we place them so you can see how space or time makes rapid fluctuations.

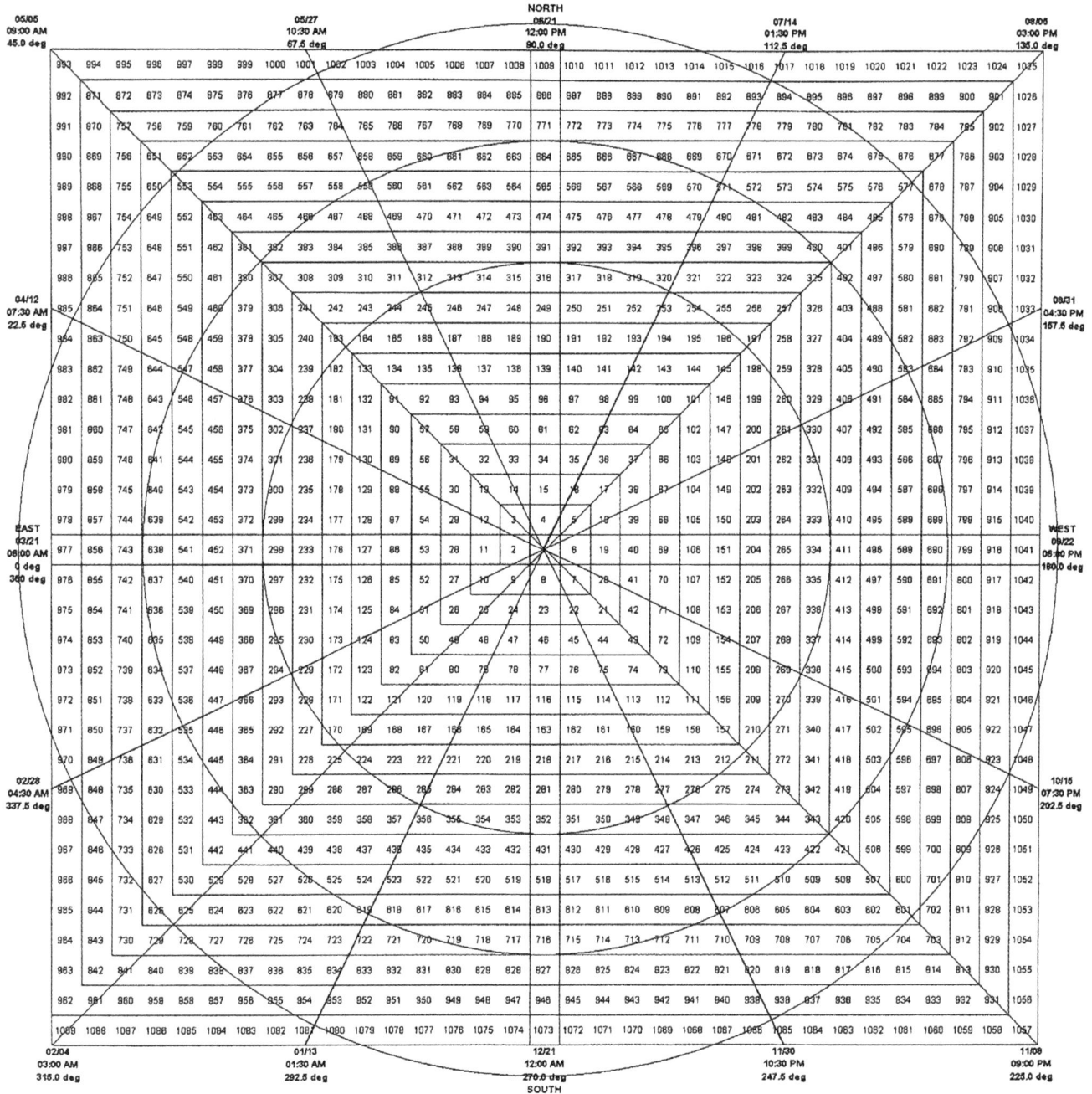

NORTH

05/05
09:00 AM
45.0 deg

05/27
10:30 AM
67.5 deg

06/21
12:00 PM
90.0 deg

07/14
01:30 PM
112.5 deg

08/05
03:00 PM
135.0 deg

04/12
07:30 AM
22.5 deg

08/31
04:30 PM
157.5 deg

EAST
03/21
06:00 AM
0 deg
360 deg

WEST
09/22
06:00 PM
180.0 deg

02/28
04:30 AM
337.5 deg

10/15
07:30 PM
202.5 deg

02/04
03:00 AM
315.0 deg

01/13
01:30 AM
292.5 deg

12/21
12:00 AM
270.0 deg

11/30
10:30 PM
247.5 deg

11/08
09:00 PM
225.0 deg

SOUTH

112

WHAT GANN SAID ABOUT THE SQUARE OF NINE

Study the SQUARE OF NINE very carefully in connection with the MASTER TWELVE and 360° CIRCLE CHART.

SIX SQUARES OF NINE

1	2	3	4	5	6	7	8	9	1	2	3	4	5	6	7	8	9	1	2	3	4	5	6	7	8	9
9	18	27	36	45	54	63	72	81	90	99	108	117	126	135	144	153	162	171	180	189	198	207	216	225	234	243
8	17	26	35	44	53	62	71	80	89	98	107	116	125	134	143	152	161	170	179	188	197	206	215	224	233	242
7	16	25	34	43	52	61	70	79	88	97	106	115	124	133	142	151	160	169	178	187	196	205	214	223	232	241
6	15	24	33	42	51	60	69	78	87	96	105	114	123	132	141	150	159	168	177	186	195	204	213	222	231	240
5	14	23	32	41	50	59	68	77	86	95	104	113	122	131	140	149	158	167	176	185	194	203	212	221	230	239
4	13	22	31	40	49	58	67	76	85	94	103	112	121	130	139	148	157	166	175	184	193	202	211	220	229	238
3	12	21	30	39	48	57	66	75	84	93	102	111	120	129	138	147	156	165	174	183	192	201	210	219	228	237
2	11	20	29	38	47	56	65	74	83	92	101	110	119	128	137	146	155	164	173	182	191	200	209	218	227	236
1	10	19	28	37	46	55	64	73	82	91	100	109	118	127	136	145	154	163	172	181	190	199	208	217	226	235
1	2	3	4	5	6	7	8	9	1	2	3	4	5	6	7	8	9	1	2	3	4	5	6	7	8	9
252	261	270	279	288	297	306	315	324	333	342	351	360	369	378	387	396	405	414	423	432	441	450	459	468	477	486
251	260	269	278	287	296	305	314	323	332	341	350	359	368	377	386	395	404	413	422	431	440	449	458	467	476	485
250	259	268	277	286	295	304	313	322	331	340	349	358	367	376	385	394	403	412	421	430	439	448	457	466	475	484
249	258	267	276	285	294	303	312	321	330	339	348	357	366	375	384	393	402	411	420	429	438	447	456	465	474	483
248	257	266	275	284	293	302	311	320	329	338	347	356	365	374	383	392	401	410	419	428	437	446	455	464	473	482
247	256	265	274	283	292	301	310	319	328	337	346	355	364	373	382	391	400	409	418	427	436	445	454	463	472	481
246	255	264	273	282	291	300	309	318	327	336	345	354	363	372	381	390	399	408	417	426	435	444	453	462	471	480
245	254	263	272	281	290	299	308	317	326	335	344	353	362	371	380	389	398	407	416	425	434	443	452	461	470	479
244	253	262	271	280	289	298	307	316	325	334	343	352	361	370	379	388	397	406	415	424	433	442	451	460	469	478

We are sending you six permanent charts, each containing 81 numbers. The first Square of Nine runs from 1 to 81. Everything must have a bottom, top, and four sides to be a square or cube. The first square running up to 81 is the bottom, base, floor or beginning point. Squares #2, 3, 4 and 5 are the four sides, which are equal and contain 81 numbers. The sixth Square of Nine is the top and means that it is times times as referred to in the Bible, or a thing reproducing itself by being multiplied by itself. Nine times nine equals 81 and six times 81 equals 486. We can also use 9 times 81, which would equal 729 (the 9th square of nine).

The number 5 is the most important number of the digits because it is the balance or main center. There are four numbers on each side of it. Note how it is shown as the balancing or center number in the Square of Nine.

3	6	9
2	5	8
1	4	7

We square the circle by beginning at 1 in the center and going around until we reach 360. Note that the Square of Nine comes out at 361. The reason for this is it is 19 times and the 1 to begin with and one over 360 represents the beginning and ending points. 361 is a transition point and begins at the next circle. Should we leave the first space blank or make it "0", then we would come out at 360. Everything in mathematics must prove. You can begin at the center and work out, or begin at the outer rim and work in to the center. Begin at the left and work right to the center or to the outer rim or square.

Note the Square of Nine or the square of the circle where we begin with 1 and run up the side of the column to 19, then continue to go across until we have made 19 columns, again the square of 19 by 19. Note how this proves up the circle. One-half of the circle is 180°. Note that in the grand-center, where all angles from the four corners and from the east, west, north and south reach gravity center, number 181 appears, showing that this point we are crossing the equator or gravity center and are starting on the other half of the circle.

Study the human body in every way and you will find that it is the work of a master mind, and when once you know yourself and know your body, you will know the law and will understand all there is to know. Remember there is a source of all supply, and that you have within you the power to know all there is to know, but you must work hard, seek and you shall find. This ends everything that Gann said about the Square of Nine.

The reference to the Square of Nine or the "Square of the Circle" above is not referring to the Square of Nine chart. Many students have mistaken this to be the same chart. The chart below is a copy of what Gann was talking about. It is simply a square of 19 by 19.

1	2	3	4	5	6	7	8	9	10	11	12	13	14	15	16	17	18	19
19	38	57	76	95	114	133	152	171	190	209	228	247	266	285	304	323	342	361
18	37	56	75	94	113	132	151	170	189	208	227	246	265	284	303	322	341	360
17	36	55	74	93	112	131	150	169	188	207	226	245	264	283	302	321	340	359
16	35	54	73	92	111	130	149	168	187	206	225	244	263	282	301	320	339	358
15	34	53	72	91	110	129	148	167	186	205	224	243	262	281	300	319	338	357
14	33	52	71	90	109	128	147	166	185	204	223	242	261	280	299	318	337	356
13	32	51	70	89	108	127	146	165	184	203	222	241	260	279	298	317	336	355
12	31	50	69	88	107	126	145	164	183	202	221	240	259	278	297	316	335	354
11	30	49	68	87	106	125	144	163	182	201	220	239	258	277	296	315	334	353
10	29	48	67	86	105	124	143	162	181	200	219	238	257	276	295	314	333	352
9	28	47	66	85	104	123	142	161	180	199	218	237	256	275	294	313	332	351
8	27	46	65	84	103	122	141	160	179	198	217	236	255	274	293	312	331	350
7	26	45	64	83	102	121	140	159	178	197	216	235	254	273	292	311	330	349
6	25	44	63	82	101	120	139	158	177	196	215	234	253	272	291	310	329	348
5	24	43	62	81	100	119	138	157	176	195	214	233	252	271	290	309	328	347
4	23	42	61	80	99	118	137	156	175	194	213	232	251	270	289	308	327	346
3	22	41	60	79	98	117	136	155	174	193	212	231	250	269	288	307	326	345
2	21	40	59	78	97	116	135	154	173	192	211	230	249	268	287	306	325	344
1	20	39	58	77	96	115	134	153	172	191	210	229	248	267	286	305	324	343

As you can see, Gann did not leave any instructions on how to use the Square of Nine. One of the unique properties of the number 9 that Gann was spotlighting, when he said that "we use 9 numbers to measure everything and we can not go past nine without starting over and using the zero" has to do with the mathematics of multiplying or adding 9 to any number. If you add "9" to any number, the resulting number will reduce back to the original single digit number through addition. For example, 1 + 9 = 10 and 1 + 0 = 1 again. 2 + 9 = 11 and 1 + 1 = 2. 3 + 9 = 12 and 1 + 2 = 3 and so on. If you multiply any number by "9," the resulting product will reduce to a "9" through addition. For example, 9 x 5 = 45 and 4 + 5 = 9. 9 X 8 = 72 and 7 + 2 = 9, 9 x 33 = 297 and 2 + 9 + 7 = 18 and 1 + 8 = 9 and so on. Also, the sum of the digits 1 through 9 = 45, i.e., 1+2+3+4+5+6+7+8+9 = 45 and this also reduces to a "9" because 4 + 5 = 9. This is why Gann says that the number 9 will measure everything.

As one can easily see, Gann considered these divisions of a circle very important for measuring time when looking for future changes in trend.

DISCLAIMER

The purpose of this material is to provide and share ideas that I have discovered in the writings of W.D. Gann. I have made every effort in the presentation of the information within this course to reproduce techniques that I believe W.D. Gann used based upon my personal interpretation of all available materials. The information should not be used or taken as trading advice including future projections. Neither I nor anyone else involved in the production of this material, will be liable for any loss, damage or liability directly or indirectly caused by the usage of this material. The data used is believed to be from reliable sources but cannot be guaranteed. There is considerable risk of loss in futures, stock and options trading. You should only use risk capital in all such endeavors. Past performance is not indicative of future results.

INTRODUCTION TO THE SQUARE OF 9

Over the years, many market enthusiasts have become familiar with the remarkable forecasting and trading record of W.D. Gann. Many so called "Gann experts" have tried to figure out how Gann was able to trade with such a high accuracy rate and make so many incredible market predictions. For example, the Ticker and Investment Digest, volume 5, number 2, December 1909, shows that W.D. Gann took a total of 286 trades in the presence of William E. Gilley, of which 264 were profitable winning trades. His success rate was an amazing 92%. In this 25-day period, which the article covers, Gann was able to double his initial capital ten times, for a gain of 1000% on his margin.

It has also been reported that Gann carried a miniature version of the Square of Nine with him into the trading pits during his most successfully recorded trades. The source of this information is Mr. Renato Alghini, an associate of Gann's for nearly six years. Gann believed that every top and every bottom in the markets had a mathematical counterpart in both price and time. He quoted Faraday saying, *"There is no chance in nature, because mathematical principles of the highest order lie at the foundation of all things. There is nothing in the universe but mathematical points of force."*

The true origin of the Square of Nine is unknown. It is believed that Gann discovered it in either Egypt or India and that it is of some ancient origin. I have talked with some friends of mine who are from India and they had never seen anything like the Square of Nine. One suggested that it may have something to do with a type of Vedic astrology but he was only guessing. My personal belief is that it probably came from Egypt because the Temple of Luxor incorporates the Square of Nine in its architecture. However, this is only an opinion. The truth is that nobody knows for certain where it came from. Maybe Gann invented the thing himself? The only thing we know for certain is that Gann used the Square of Nine and considered it very valuable.

In his Egg course, Gann describes the square of four as the "even square" because it is the square of the first even number, i.e., the square of 2. The first odd square would be "1" but this does not produce a number greater than itself because 1 x 1 = 1. The first odd square greater than itself is "9" or 3 squared. Gann said, *"We use the square of odd and even numbers to get not only the proof of market movements, but the cause."*

WHAT IS THE SQUARE OF 9
(KEEP A SQUARE OF NINE CHART IN FRONT OF YOU WHILE READING)

The Square of Nine is basically a spiral of numbers starting with the number one in the center (or apex of the Great Pyramid) with the number 2 immediately to the left. The rest of the numbers spiral around the center in a clockwise fashion to the number 9, which completes the first cycle of numbers around the center. 10 through 25 completes the 2nd cycle, 26 through 49 completes the 3rd, etc. The square is divided into eight 45-degree angles. On the cover page, I have provided a copy of the Square of Nine. You also have a large Square of Nine chart (for Microsoft Excel) included with this course that you can refer to as well.

The numbers that run through the center in the shape of a "+" sign are the cardinal numbers. The numbers that run through the center in the shape of a "X" are the corner numbers. In the first cycle around the center, there is one digit separating each 45-degree angle. In cycle number two (10 to 25) there are two digits separating each 45-degree angle. In cycle three (26 to 49), there are three digits separating each 45-degree angle. In cycle 1000, there would be 1000 digits or cells separating each 45-degree angle. Technically, the number "1" in the center is a complete cycle and would therefore be cycle #1, but there is a nice simple mathematical relationship to the cycle number and the difference between numerical values of the 45-degree angles when you count the Square of Nine numbers in this manner.

To fully appreciate the Square of Nine in terms of its geometric origins, take a look at the large chart of the Square of Nine that is included with this course. Try to visualize it as a pyramid. At the very top or apex of the pyramid is the number 1 and there are four equal sized triangular walls descending down to the pyramid's square base. Each block in this pyramid is given a number as you work your way down and around each level of blocks. Now if you remember, the numbers on the "+" are called the cardinal numbers. These numbers are all separated by increments of 90°, i.e., 90°, 180°, 270° and then 360°, which brings you back to the location that you started from.

The numbers on the "X," which connect the four corners of the square base, i.e., the corner numbers, also are separated by increments of 90°, giving the appearance of an Egyptian style pyramid. The cardinal "+" and corner "X" numbers divide the square base of the pyramid into 8 equal divisions of 45°, hence its other popular name "The Octagon Chart" (Octa meaning eight).

If you look at the Square of Nine chart, you will also see circles or rings, which have been drawn around certain squares. The last circle has calendar dates that revolve in a clockwise fashion around the square base, which starts from the date March 21st. This is the vernal equinox, when the sun is at 0° Aries, also known as the 1st day of spring and represents the beginning of the natural year. Actually, the sun does not move, it only appears to be at 0° Aries. In reality, the Earth, which revolves around the sun, as one of its many satellites, is at the opposite sign of 0° Libra. We will get into more detail of the Zodiac in a little while.

The first innermost circle has a radius that runs from the center down to the cardinal "+" number 352. This number appears on the same cycle or level of pyramid blocks as the number 360, which is 8 blocks to the left of 352. As you know, there are 360° in a circle and this is why Gann has a ring around this particular square. The next circle or ring from the center has a radius that runs down to the cardinal "+" number 716. This number appears on the same cycle or level as the number 720, which equals 2 times 360 and is the reason the second ring is located here. The third ring runs through the number 1080, which equals 3 times 360, etc. Gann set his third ring radius four blocks below thd 2nd ring which was the amount of space or blocks between ring 2 and ring 1.

The reason for constructing a chart like this is based upon the hypothesis that each positive whole number, i.e., the regular counting numbers (1, 2, 3, 4, 5, etc.) all correspond to some specific angle of a circle between 0° and 360°. Pythagoras, one of history's greatest mathematicians and philosophers, said, *"Units in a circle or in a square are related to each other in terms of space and time at specific points."* The Square of Nine is unique in that it achieves the ancient practice of squaring the circle and is often called The Pythagorean Cube.

Notice how the square completes at the corner number "X" 361 on one of the 45° angles of the Square of Nine, the 315° angle. If you started with a zero in the center, it would have came out exactly at 360. Notice how the second ring from the center has a radius that runs through 720, (2 x 360) as stated earlier, but also perfectly inscribes the 361 block (square base) within its radius.

Now if you look at the number "4" on the chart and follow an angle of 45° going up to the top right hand corner, you get the number series: 4, 16, 36, 64, 100, 144, etc. These numbers are all squares of even numbers, i.e., 2 x 2, 4 x 4, 6 x 6, 8 x 8, 10 x 10, 12 x 12, etc., respectively. If you look at the number "1" on the chart and follow an angle of 45° going down to the bottom left hand corner, you get the number series 1, 9, 25, 49, 81, 121, 169, etc. These numbers are all squares of odd numbers, i.e., 1 x 1, 3 x 3, 5 x 5, 7 x 7, 9 x 9, 11 x 11, 13 x 13, etc., respectively. Gann said, *"We use the square of odd and even numbers to get, not only the proof of market movements, but the cause."*

This particular arrangement of numbers on the Square of Nine creates a very unique square root relationship with all the other numbers on the Square ofNine chart. My friend Michael S. Jenkins illustrates some interesting square root trading techniques utilizing the Fibonacci ratios with the Square of Nine in his *Stock Trading Course* and his book *Chart Reading for Professional Traders,* published by Traders Press, Inc. http://www.traderspress.com.

NAVIGATING WITH THE SQUARE OF NINE

If you want to move in cycles of 360° around the Gann Wheel, you take the number you are interested in, such as the all time high or low price, take the square root of the number, then add or subtract 2 from the root and re-square the result.

Example: Let's say that we are interested in the price 664, which is in the vertical column straight up from the center, the square root is 25.768 + 2 = 27.768^2 = 771, which is the number directly above 664 or one full 360° degree cycle out from center. If we subtracted 2 from the square-root and re-squared the number (25.768-2 = 23.768^2 = 564.9) we would get 565, which is directly below 664 or one full 360° degree cycle in towards center. The reason this works is that the squares of the "even" and "odd" numbers line up on a straight line from the apex or main center square of the chart. It's also mathematically simple to observe that all odd numbers are separated by units of 2, such as 1, 3, 5, 7, 9, 11, etc. The same is obviously true for all the even numbers: 2, 4, 6, 8, 10, 12, etc. This is why adding or subtracting "2" to the square root of a number, then re-squaring the sum is equivalent to a 360° cycle on the Square of Nine. Another mathematical proof is that $\frac{1}{2}$ a circle is 180° and we can see that the squares of the "even numbers" line up on the opposite side of the Square of Nine to the squares of the "odd numbers." We learned that adding "2" to the square root of a number, re-squared was equal to 360°. Therefore, adding "1" to the square root of a number, re-squared would have to be equivalent to 180°, because $\frac{1}{2}$ of 2 equals "1" and $\frac{1}{2}$ of 360° equals 180°. If we wanted 90° relationships, we would add or subtract 0.5 to the square root, then re-square because 90° is $\frac{1}{4}$ of 360° and $\frac{1}{2}$ of 180° and 0.5 is $\frac{1}{4}$ of "2" and $\frac{1}{2}$ of "1," etc.

Gann said, *"We use three figures in geometry: the circle, the square and the triangle. We get the square and triangle points of a circle to determine points of time, price and space resistance; we use the circle of 360° to measure time and price."* Gann's Emblem was a square and triangle inside a circle. Incorporating the Gann Emblem with the square root technique allows us to calculate coordinates or numbers that are conjunct (360° = +/- 2 from the root), opposition (180° = +/- 1 from root #), trine (120° = +/-0.666) & (240° = +/- 1.333) or square (90° = +/- 0.5) and (270° = +/- 1.5). This technique is extremely fast and useful for finding coordinate cells (pyramid blocks) on the Square of Nine that have a geometrical relationship to a previous position on the chart. Gann also used these same geometric relationships to divide the outer calendar circle that circumscribes the Square of Nine. Gann basically divided the Earth's 360-degree orbital cycle around the sun into quarters and thirds to find geometric relationships in terms of time as well. The quarter divisions are 90, 180, 270 and 360 degrees. The one-third divisions are 120, 240 and 360 degrees. Because the Earth on average will move about 1-degree per day, Gann used these numbers as calendar days, which he added to the dates of previous tops and bottoms to find dates in the future that had a mathematical or geometrical relationship to past highs and lows.

Gann believed that the numbers that connected the square base of the pyramid (the 4 diagonal corners "X" of the square, i.e., corner #'s) to the "main center" and also the numbers that ran straight vertical and horizontal from the "gravity center" in the form of a cross (cardinal numbers "+") were very important in balancing "price and time" on the wheel. He was basically looking for the astronomical longitudes of the sun or Earth to balance with price on these key angles. Remember, Pythagoras said, *"Units in a circle or in a square are related to each other in terms of space and time at specific points."* Gann often quoted the Bible, Emerson, Pythagoras and Faraday to name a few. Basically, he was pointing the reader of his works to clues that would allow them to unlock the code of his writing style.

Because many readers of this course are probably not familiar with using longitude, planets or the divisions of the zodiac, we give the following explanation. The zodiac is an imaginary circle based upon the apparent path of the sun, as it appears to rise and set in circular motion against the backdrop of the constellations or stars. This is similar to the equator being an imaginary circle at the center of our planet. We measure longitude on earth in degrees and minutes of the imaginary circle (equator) west of Greenwich England, which represents the 0° starting point of the circle.

The zodiac is measured in a similar fashion. We measure the heavens in an imaginary 360° circle called the zodiac. It is measured in degrees and minutes from 0° Aries, which is the location of the sun at the vernal equinox, i.e., spring. The zodiac is divided into 12 equal divisions of 30° each. Aries runs from 0° to 30°, Taurus runs 30° to 60°, Gemini runs 60° to 90°, Cancer runs 90° to 120°, Leo runs 120° to 150°, Virgo runs 150° to 180°, Libra runs 180° to 210°, Scorpio runs 210° to 240°, Sagittarius runs 240° to 270°, Capricorn runs 270° to 300°, Aquarius runs 300° to 330° and Pisces runs 330° to 360° or back to 0° Aries.

If you look at your Square of Nine chart, you will see the date March 21st, at the 9 o'clock position. This is the date that the sun appears to be at 0° Aries, beginning the natural year. *"Lamb of God, you take away the sins of the world,"* the season of spring takes away the sins of winter. Moving in a clockwise fashion to the top left corner, you find the date May 5th a.k.a. Cinco de Mayo. On this day, the sun appears to be at 15° of Taurus. The next date at the top of the chart in the 12 o'clock position is June 21st. This is the summer solstice, which is the longest day of the year for the northern hemisphere. The sun appears to be at 0° of Cancer on this date. The top right hand corner of the chart has the date August 5th. The sun now appears to be at 15° of Leo. On the right-hand side of the chart at the 3 o'clock position is the date September 22nd. This date is the Autumnal Equinox, which begins the season of fall, when the sun literally begins to fall below the Earth's equator, i.e., declination. The sun appears to be at 0° of Libra on this day. Moving down to the bottom right hand corner of the chart, the date November 6th appears, the sun is now at 15° of Scorpio. The bottom of the chart at the 6 o'clock position has the date December 21st, which is the winter solstice. This is the shortest day of the year in the northern hemisphere, marking the season of winter. The sun appears to be at 0° Capricorn on this date. Moving over to the bottom left hand corner, we see the date February 4th where the sun appears to be at 15° Aquarius on this date. Moving up takes us full circle, or back to 0° Aries or March 21st.

Notice how the four seasons are aligned with the cardinal "+" numbers of the Square ofNine. This relationship allows the user of the Square of Nine chart to correlate the sun or other planetary longitudes to the numbers on the chart with a measure of time. The user can also quickly locate dates that are square (90° and 270°), opposite (180°) or trine (120° and 240°) to a past date based upon the apparent longitude of the sun.

The fundamental philosophy that Gann was spotlighting was simply that March 21st was the annual reincarnation of life's new birth. By June 21st at the summer solstice, intense new growth has taken place in the animal, plant and insect kingdoms. By the autumnal equinox on September 22nd, the peak of vitality and fruition was being reached and the life cycle was ebbing towards old age as the new dominant cycle. By December 21st, at the winter solstice, the life cycle is in a frozen state of suspended animation as animals hibernate, plants appear dead and the suspended seeds of the next generation of new life lie waiting for the resurrection of the life-giving rays of the sun, at the vernal equinox. The Bible plainly states that there is a time and a season for everything under the sun and there are no exceptions to this cyclic rule. The Earth revolves around the sun in 365 days or one year. This completes its journey through the 12 signs of the zodiac or 360° orbit. Gann instructed his students to read the Bible at least 3 times because his greatest discoveries were found in the Bible. He considered it the greatest scientific book ever written.

Copy this image to Mylar or any other type of transparent material to be used as an overlay with the Square of Nine chart.

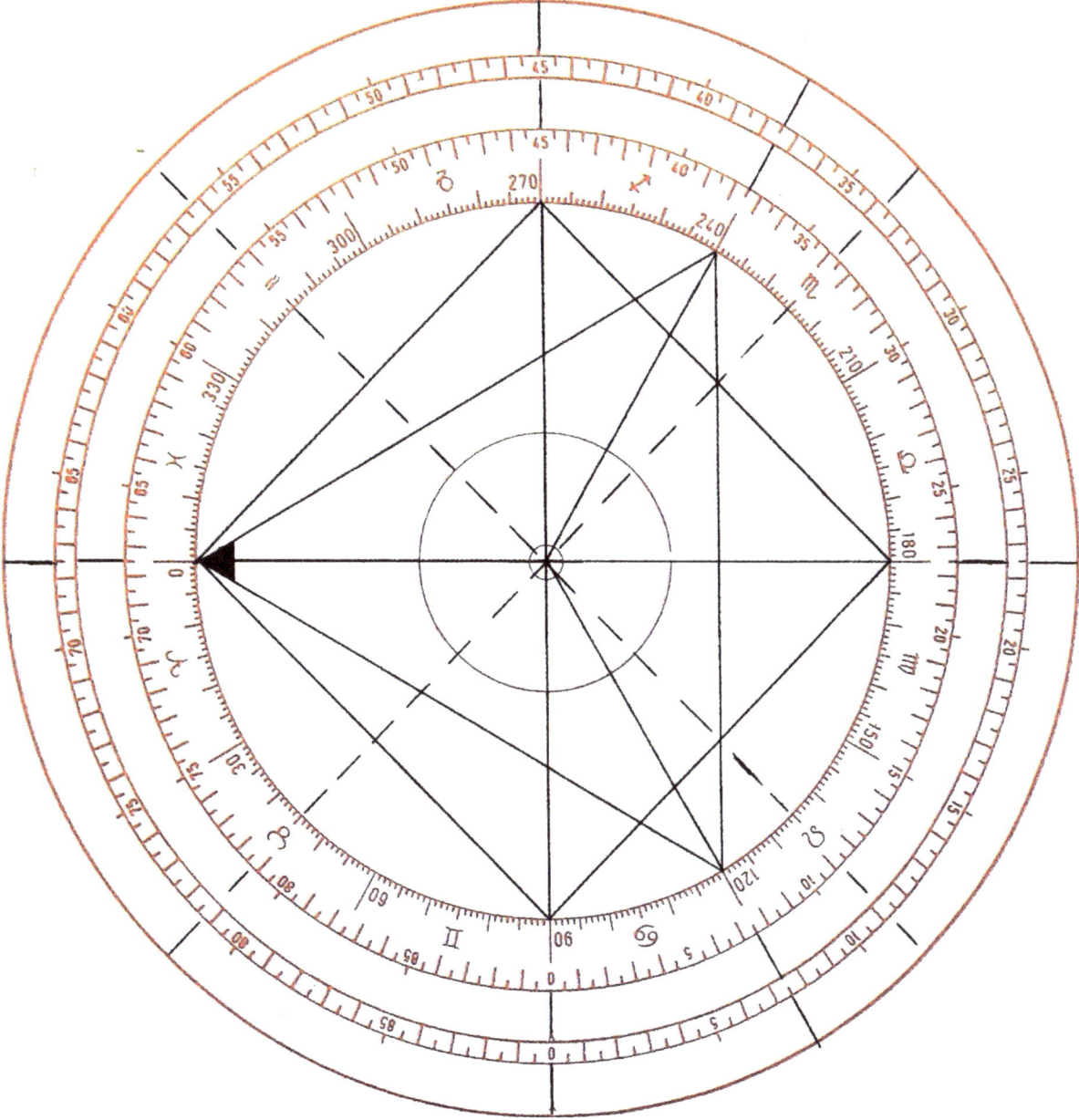

PRICE TARGETS FOR SUPPORT AND RESISTANCE

Calculating price targets for support and resistance is a very simple task. As stated earlier, the Square of Nine's number arrangement is such that the numbers have a simple square root relationship to other numbers on the wheel. I will illustrate the basic square root calculations, but if you prefer to use the chart itself, I have provided a clear plastic overlay that you can place on top of the Square of Nine to see the geometric relationships. Just simply place the 0° line of the overlay so that it runs across your price number and also through the center "1" of the Square of Nine chart. For example, if the market you were trading had an important low at $2.23, you would place the overlay so that the 0° line connects the number 223 to the center "1." Now you can quickly see the other numbers that are square (90° & 270°), trine (120° & 240°), semisquare (45°, 315°), sextile (60° & 330°), etc. As discussed earlier, these "pressure points" can be mathematically calculated using simple addition or subtraction to the square root of the price. To calculate what a geometric relationship is equal to as a square root increment, we simply divide the number by 180 as illustrated below. The numbers in bold print are considered more important for support or resistance than the numbers in regular print.

45° = 45/180 = 0.25	**270° = 270/180 = 1.50**
60° = 60/180 = 0.333	300° = 300/180 = 1.666
90° = 90/180 = 0.50	**315° = 315/180 = 1.75**
120° = 120/180 = 0.666	**360° = 360/180 = 2.0**
135° = 135/180 = 0.75	
180° = 180/180 = 1.0	
225° = 225/180 = 1.25	
240° = 240/180 = 1.333	

Taking our example of $2.23, we would first treat this as the number 223 on the Square of Nine. This is because the Square of Nine tends to work better when you float the decimal point on prices, making all prices either a three or a four-digit whole number. We will first assume that 223 is a low price, and that we want to calculate future resistance levels. We simply take the square root of 223, which is 14. 93 and add the increments we calculated on the previous page to this root number (14.93) and re-square.

45° = 45/180 = 0.25	**14.93 + 0.25 = 15.18^2 = $230.43**
60° = 60/180 = 0.333	14.93 + 0.333= 15.263^2 = $232.96
90° = 90/180 = 0.50	**14.93 + 0.50 = 15.43^2 = $238.08**
120° = 120/180 = 0.666	14.93 + 0.666 = 15.59^2 = $243.23
135° = 135/180 = 0.75	**14.93 + 0.75 = 15.68^2 = $245.86**
180° = 180/180 = 1.0	**14.93 + 1.0 = 15.93^2 = $253.76**
225° = 225/180 = 1.25	**14.93 + 1.25 = 16.18^2 = $261.79**
240° = 240/180 = 1.333	14.93 + 1.333 = 16.263^2 =$264.48
270° = 270/180 = 1.50	**14.93 + 1.5 = 16.43^2 =$269.94**
300° = 300/180 = 1.666	14.93 + 1.666 = 16.596^2 =$275.42
315° = 315/180 = 1.75	**14.93 + 1.75 = 16.68^2 =$278.22**
360° = 360/180 = 2.0	**14.93 + 2.0 = 16.93^2 = $286.62**

If our $223 price was a high instead of a low, we would have subtracted the square root increments from 14.93 and then re-squared the difference to calculate support. If you build a table of important price highs and lows, similar to what was illustrated with dates and do the above calculations, you can determine if certain prices have a cluster of geometric relationships to previous significant prices, thus making that specific price or circular degree more important for support or resistance. A simple way to do this is to determine what degree of the circle your price is on using a Square of Nine table.

GANN ANGLE PROJECTION

If you would like to project a vector or angle from a previous top or bottom, the first thing you should do is float your decimal, making your price either 3 or 4 digits, as we have done in the previous example. The next step is to calculate the dimensions of our price box. Again, we do this with the square root function. As an example, by taking the major 1361.09 price low of the S&P500 index on May 24, 2000 and going 180-degrees around the Square of Nine, we arrive at the price of 1435.87. That is to say that the square root of 1361.09 = 36.89 + 1(180°) = 37.89, re-square = 1435.87. The difference between these prices is 1435.87 — 1361.09 = 74.78 points. This means that the 1 x 1 or 45° angle moves up at the rate of 74.78 points in 37 days. The time element is simply the square root of the May 24th low price of 1361.09 (36.89 = 37 days), which comes out as June 30th. The other important Gann angles or vectors are the 2 x 1, 3 x 1, 4 x 1, 8 x 1, 1 x 2, 1 x 3, 1 x 4 and 1 x 8. The 2 x 1 angle will simply advance at double the rate of the 1 x 1 angle. So in this example, the 2 x 1 angle advances (74.78 x 2) 149.56 points in 37 days. The 1 x 2 angle would move up at $^1/_2$ the rate of the 1 x 1 angle or 37.39 points in 37 days, etc. If you will take the time to investigate this technique, you will immediately see how accurate this forecasting technique really is. It is also important to note that some markets will work better as a 3-digit number and others will work better as a 4-digit number (most work better as 3 digits). The choice, however, will become obvious with a little research on your behalf. If you do not have Gann's angle course, I would suggest that you review the free copy of this course that was included with your order. The Gann angle course will give you much more insight into the interpretation and application of his vectors or timing angles.

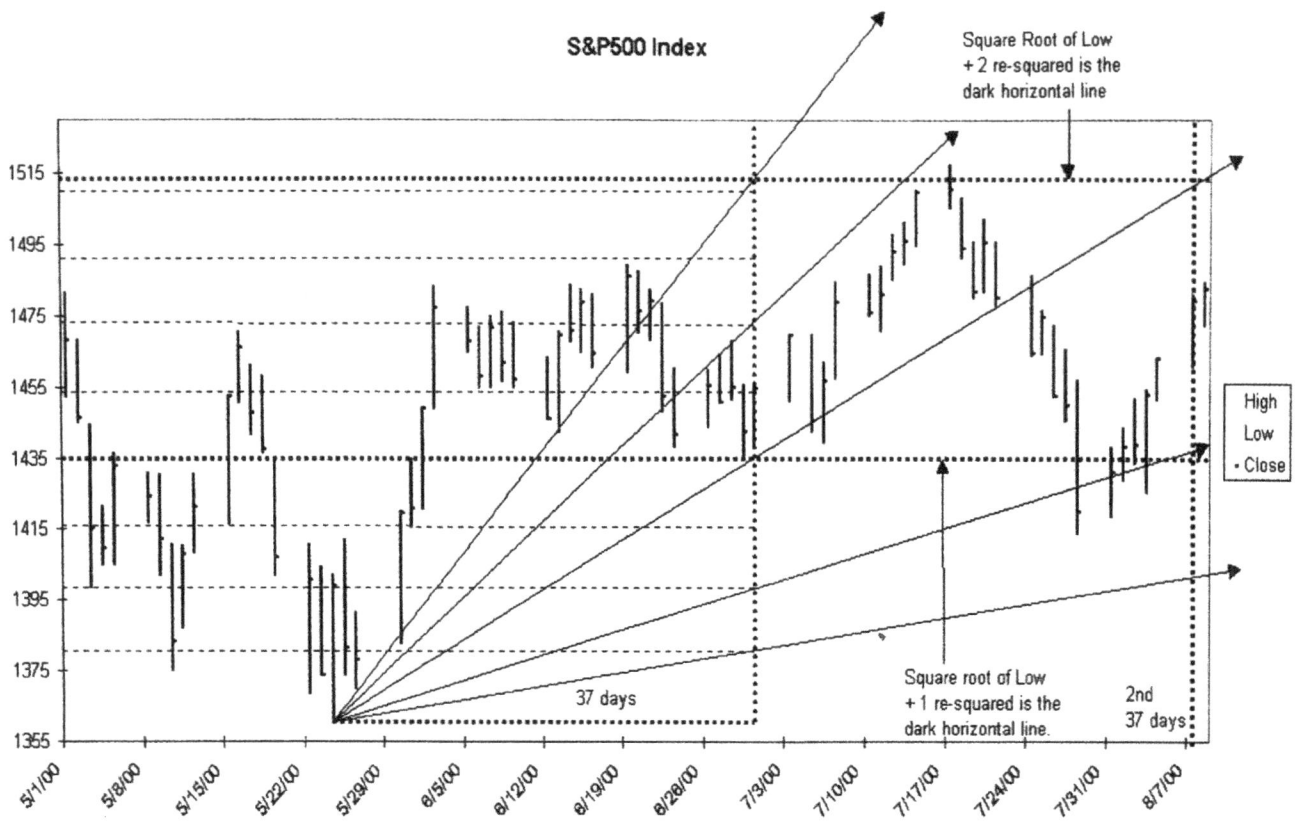

S&P500 Index

Square Root of Low + 2 re-squared is the dark horizontal line

Square root of Low + 1 re-squared is the dark horizontal line.

37 days

2nd 37 days

High
Low
- Close

The previous chart illustrates how well the S&P 500 index (as a 4-digit number) follows 180-degree root increments (adding to the root and re-square) and how to draw Gann angles from this basic technique. The highest price on the graph just slightly penetrated the 360-degree cycle of price from the May 24th low price. The decline following the highest high price at 1517.32 on July 17th, 2000 fell 240-degrees or (1.333 as a square root increment) where the market found support. The square root of the high is 38.952 so our time period is now two days longer or 39 days from this top. Adding 39 days to July 17th gives August 25, 2000, which is also the Anniversary date of the 1999 top. You should note that July 17th was the anniversary of the 1998, 1997 and 1996 tops.

126

If you look at your Square of Nine chart, you will see that the odd square of numbers (1, 9, 25, 49, etc.) line up on opposite side of the even squares (4, 16, 36, 64, etc.). This relationship is shown on the Square of Nine as the diagonal of the square and is therefore equal to a 45° angle. This is how Gann graphically illustrates the chart. The "odd squares" and "even squares" line up on a 45° angle diagonally through the main center "1." You can also prove this with the Pythagorean Theorem, which states that the sum of the squares of the sides of a right triangle is equal to the square of the hypotenuse, i.e., the diagonal. In the S&P 500 example, we advanced 74.78 points in 37 calendar days. This was equal to 26 trading days of 6 1/2 hours each session (9:30 AM to 4:00 PM). Therefore the total trading hours = (6.5 x 26) 169 hours. The square root of (169^2 + 74.78^2) equals 184.80, which is approximately 5 digits past a perfect 180. This means that the low on June 30th would have been perfectly balanced at 11:00 AM or 164 trading hours from May 24th [square root of (164^2 + 74.78^2) = 180]. The lightly dashed horizontal lines are 1/8th lines that were calculated by dividing the full 360° range by 8.

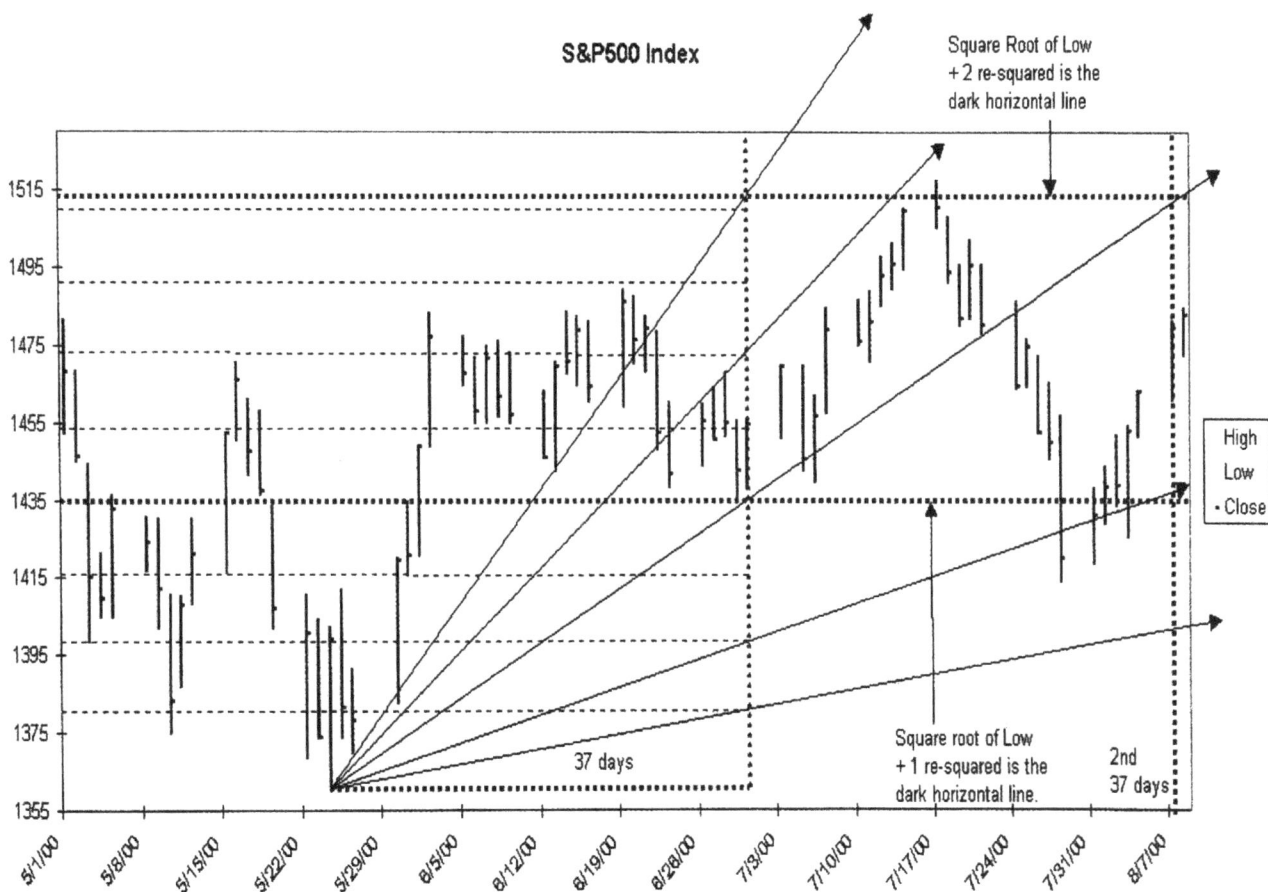

127

SPECIAL STOCK MARKET CYCLE REPORT
© January 2002

THE 18-YEAR SUPER BULL & BEAR MARKET CYCLE

There is a long-term cycle in the U.S. Equity Markets of approximately 17 to18 years. You will find that this information alone is worth more than the meager cost of this report. Anyone that takes time and observes the history of the stock averages will certainly see this trend. The problem is that most people will never look, but that is to your advantage! The Bible plainly states that there is *"an appointed time for everything, and a time for every affair under the heavens "*—Ecclesiastes 3:1. The stock market is no exception to this rule! From the stock market low in 1911, the Dow Jones Industrial Averages advanced in "super bull market" fashion to the famous market top (and crash) of 1929. The Dow Jones Industrial Average reached a price of around 386 in September of 1929. From 1929 to 1947, the market was in an 18-year "super bear market" cycle. This is when the country experienced the "Great Depression." The equity markets made their extreme low in July of 1932, which will prove to be important when I discuss the 42-year cycle. It is also important to point out that the Dow Jones Industrial Average never reached its high of 386 during this entire 18-year "super bear" cycle. From 1947 to 1965, the stock market was once again in "super bull market" mode and the Dow Jones Industrial Average broke 1000 for the very first time. From 1965, we entered another 18-year "super bear market" cycle, which would last until late 1982. Again, it is important to point out that the stock market averages basically could not exceed the preceding "super bull market" top, which was just over 1000 for the Dow Jones Industrial Average for the entire 18-year period! From 1982 to 2000, the U.S. stock markets were once again in an 18-year "super bull market" cycle. This time it resulted in the greatest bull market in history, taking the Dow Jones Industrials as high as 11,750.

From the year 2000 to approximately 2018, we will be in a "super bear market" cycle. Historically speaking, this means that the stock market averages are not likely to exceed their prior "super bull" tops. The Dow Jones Industrial Average may be the exception to the rule since it is only comprised of 30 stocks and is a little easier to manipulate. This (in my opinion) is what happened in 1974 during a " super bear market" cycle. The Dow Jones was able to get back to its prior "super bull" top before dropping 60% by December of that same year. They say, *"A picture is worth a thousand words"* and that *"seeing is believing"* The charts below should help illustrate what was just discussed in this section. It is important to note that extreme bear market bottoms come approximately every 42-43 years. As mentioned before, the lowest point reached in the "super bear market" cycle from 1929-1947 was reached in July of 1932. Forty-two years later in 1974, the stock market again reached its lowest point in a "super bear market" cycle. If this phenomenon continues, we can anticipate that an extreme stock market low will be reached in our "super bear market" cycle around the year 2016. This is two years earlier than our expected 18-year "super bear market" cycle but it will most likely be an extreme low. In other words, the 18-year cycle low in 2018 will probably be a higher low than the year 2016 just like 1982 was higher than 1974 and 1947 was higher than 1932.

The Big Picture ©

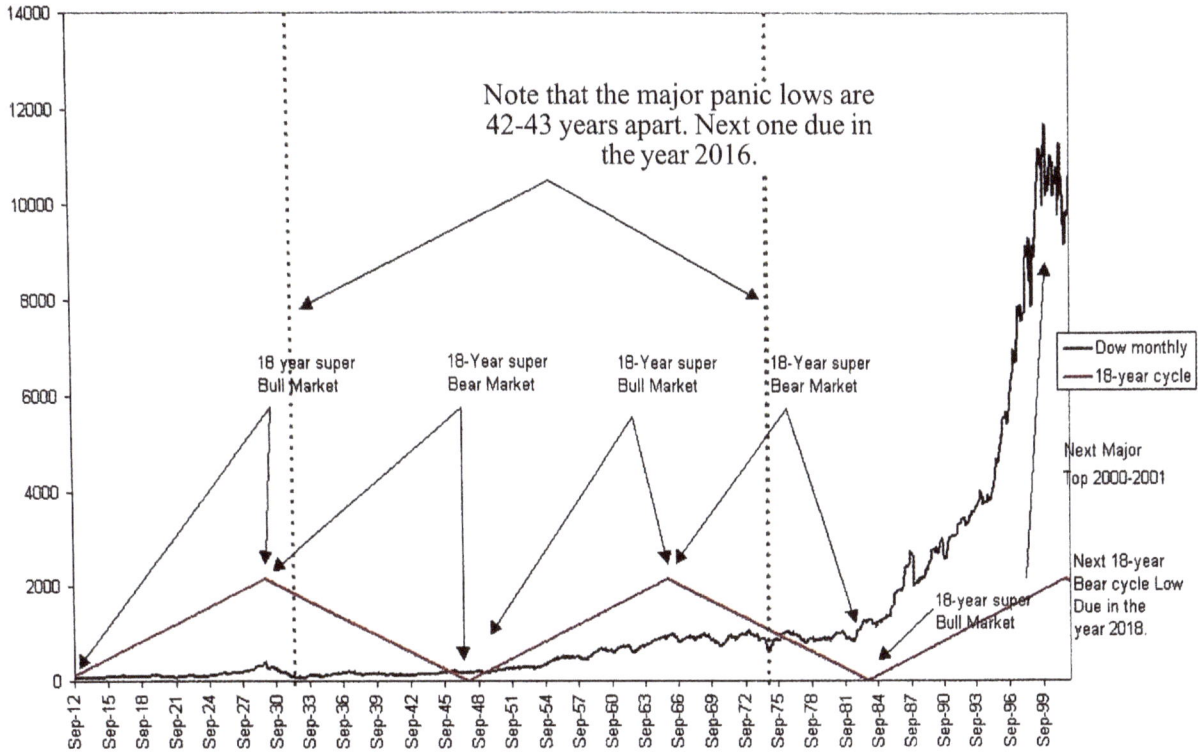

Note that the major panic lows are 42-43 years apart. Next one due in the year 2016.

18 year super Bull Market

18-Year super Bear Market

18-Year super Bull Market

18-Year super Bear Market

18-year super Bull Market

Dow monthly
18-year cycle

Next Major Top 2000-2001

Next 18-year Bear cycle Low Due in the year 2018.

A Closer Look at the Cycles
1911-1947 ©

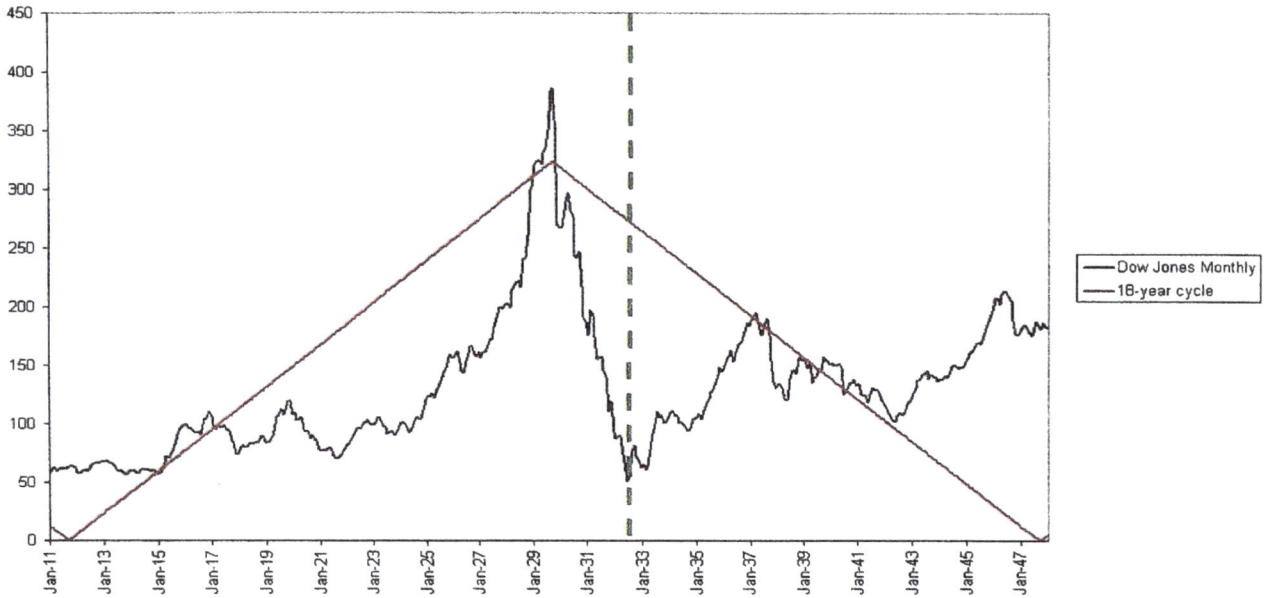

42-year low dashed vertical line.

1947-1983 ©

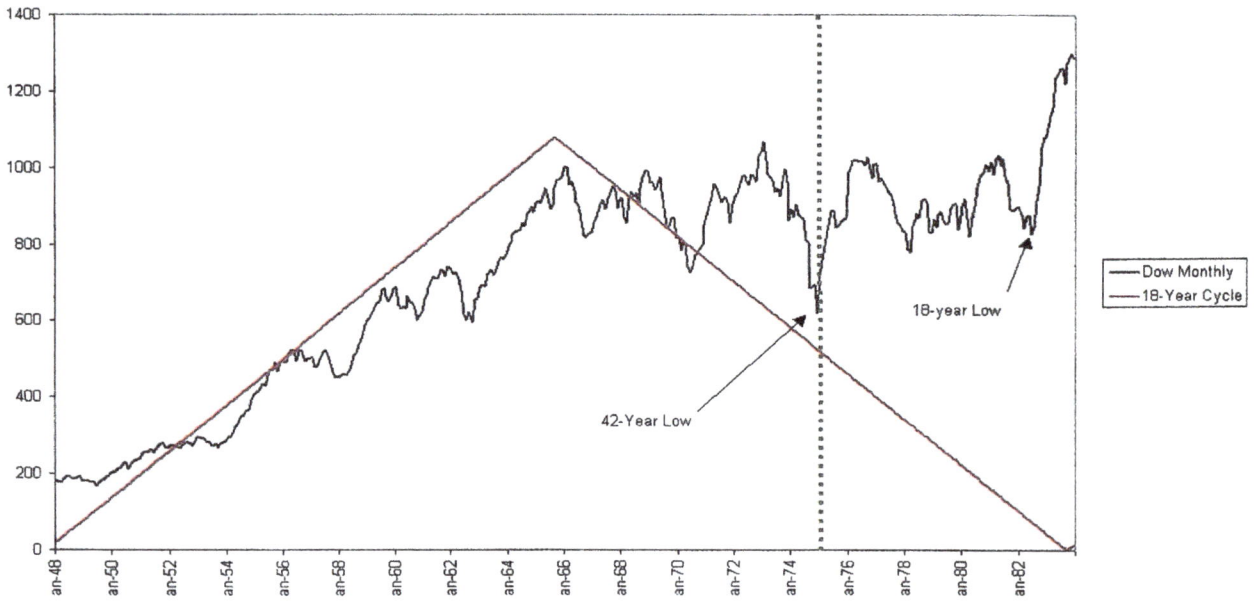

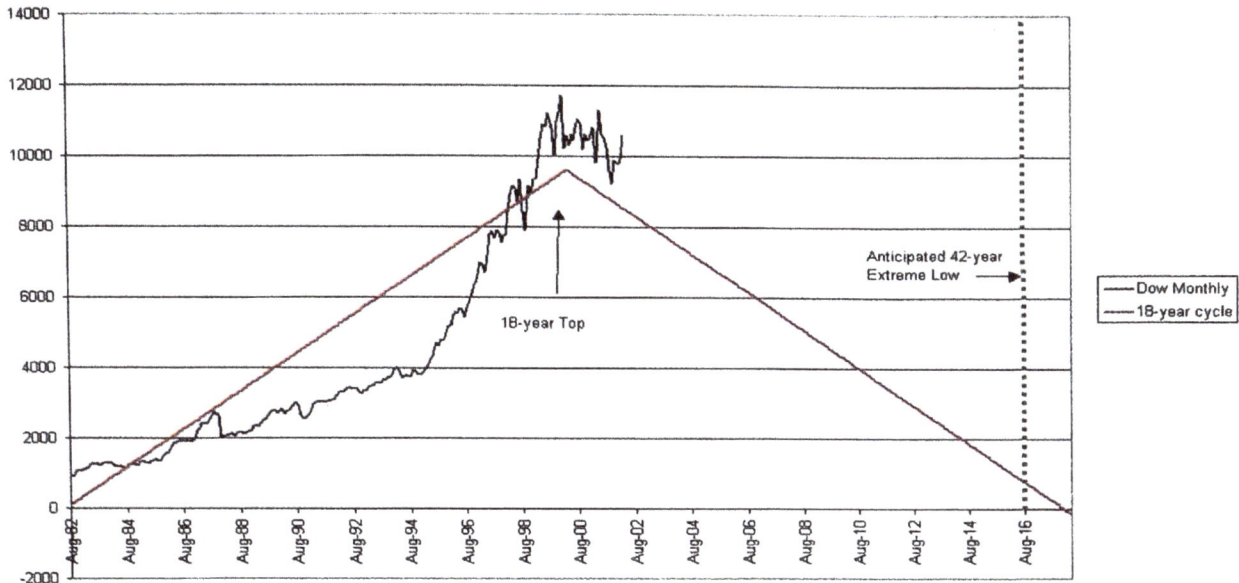

If I were investing in the stock market or mutual funds during this 18-year "super bear" cycle, I would really be interested in finding out the performance of the fund during previous bear market cycles. In other words, find the money managers that performed well in the period from 1965-1982 or 1929-1947. This is one area were a broker can actually be useful. I'm not very big on bear market mutual funds, because the long-term tendency is for the markets to advance. Remember, it has been going up for 210 years and as a result of this tendency even the 18-year "super bear" market cycles have an upward bias. In other words, the market bottoms tend to be higher than the preceding bottom within the 18-year cycle. This can clearly be seen in the above charts during the periods 1929 to 1947 and 1965 to 1982. Even though the markets could not make any significant new highs during this 18-year bear period, the lows had a tendency to be higher than preceding lows within the 18-year negative phase. When we enter the next "super bull market," you will be better educated and will not believe all the media hype. You will know how to anticipate its approximate end because it is simple addition (2018 + 18 = 2036). You will know that market corrections during this period are good buying opportunities for the remainder of the cycle. For example, in this last 18-year super bull market, the crash of 1987, the mini-crash 1990, the 1-year recession of 1994 and the corrections in 1997 and 1998 were all great buying opportunities because the 18-year cycle had not run its full course. Right now, rallies are most likely good selling opportunities as far as the general trend is concerned. I'm not saying that you have to wait until 2016 to invest in the equity markets again. I'm just saying that it's not going to be as easy as it was during the 18-year up cycle. For those that want to look for a possible planetary phenomenon that correlates to 18 years, I suggest you look at the synodic cycle of Saturn and Neptune from a heliocentric view. Each 180 degrees of synodic movement takes approximately
18 years.

THE 42-YEAR CYCLE

As briefly mentioned in the discussion of the 18-year cycle, major market lows tend to occur 42-43 years apart. The basic pattern of the 42-year cycle is different from the 18-year cycle. In the 18-year cycle, we had an even amount of up-time compared with the down-time, i.e., 18 years up and 18 years down. The 42-year cycle is a bit different. It exhibits what cycle analysists like to call right translation. This simply means that the cycle spends an unequal amount of time going up than it does going down, which is probably the "effect" of the combination of several other cycles. The basic pattern for the 42-year cycle in the Dow Jones Industrial Average is to advance for 39 years and to decline **very hard** for 3 years. One of the potential reasons that the 1929 crash was so severe was the phasing of the 18 and 42-year cycles. They both peaked in 1929 and started their down cycles leading into the worst stock market correction in our history (about a 90% drop), which was also a major contributing factor to the " Great Depression." The basic 42-year cycle shape is presented in the following graph.

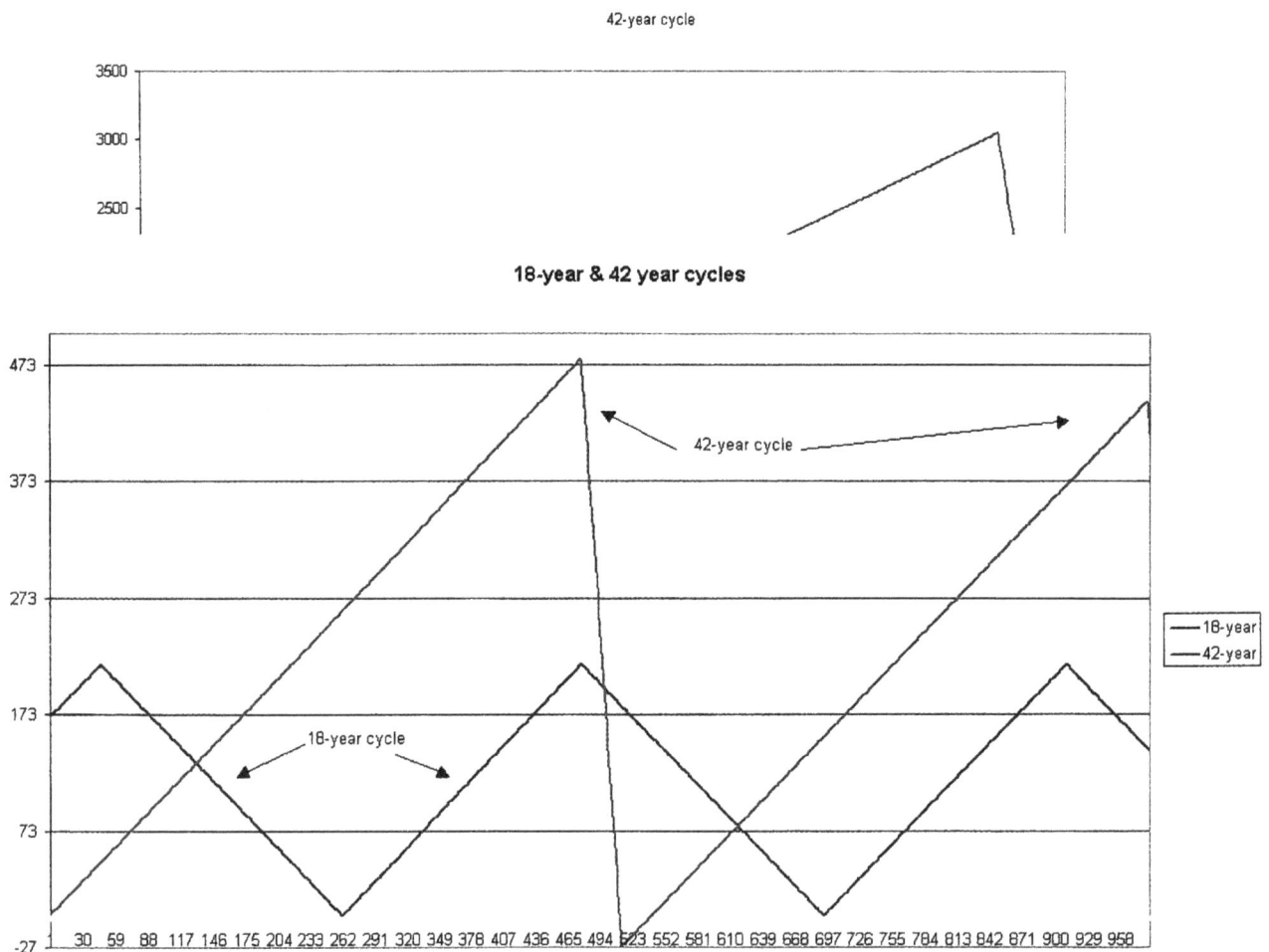

42-year cycle

18-year & 42 year cycles

When we combine this cycle with the 18-year cycle, we gain a very interesting long-term perspective of our financial markets. Combining these two powerful cycles even seems to account for the long flat periods that occur at periodic intervals in the history of our equity markets.

132

The chart below is another look at the "big picture" along with our new cyclical model of market behavior based on the combination of the 18-year and 42-year cycles. The next set of charts "zooms in" on the data to get a better perspective of how these two cycles have influenced market behavior.

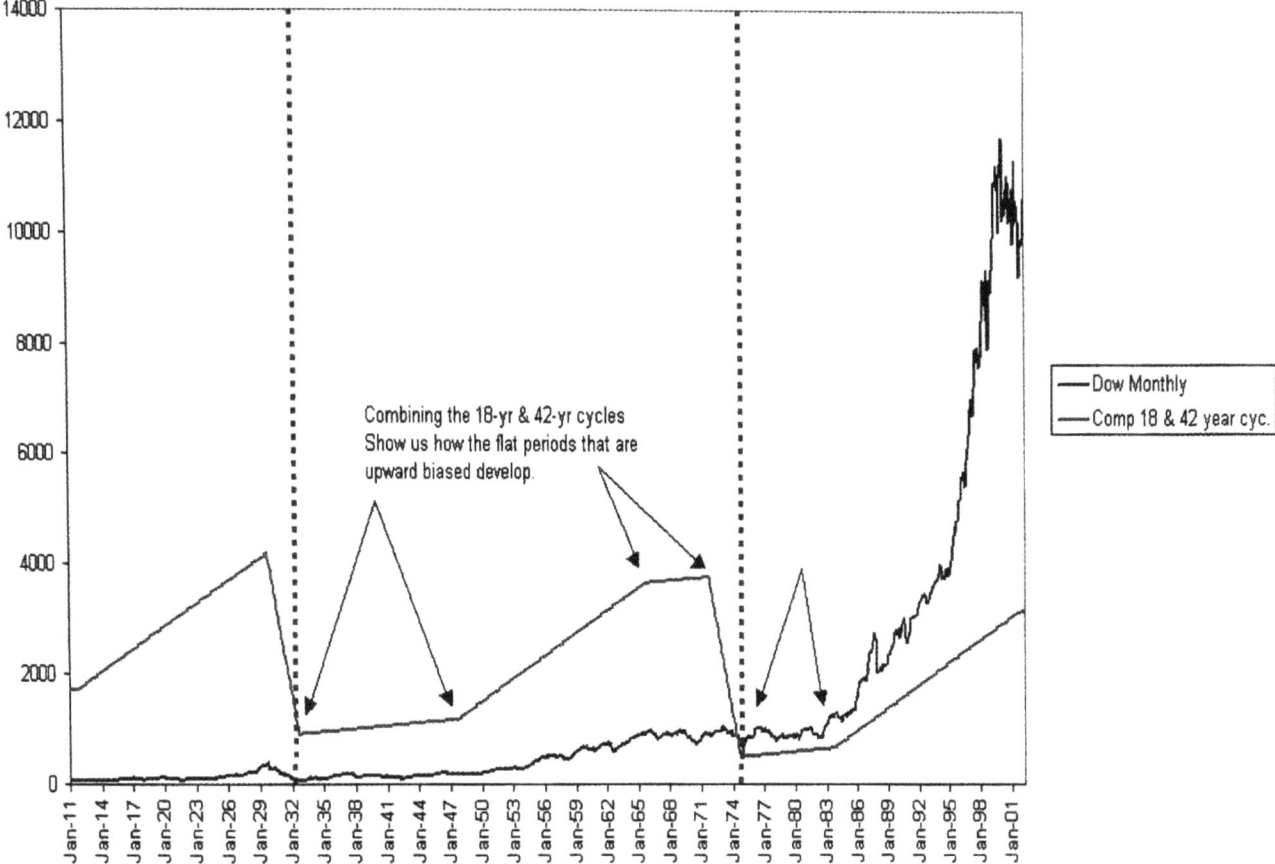

Combining the 18-yr & 42-yr cycles
Show us how the flat periods that are
upward biased develop.

Dow Monthly
Comp 18 & 42 year cyc.

Composite 1911 to 1947 ©

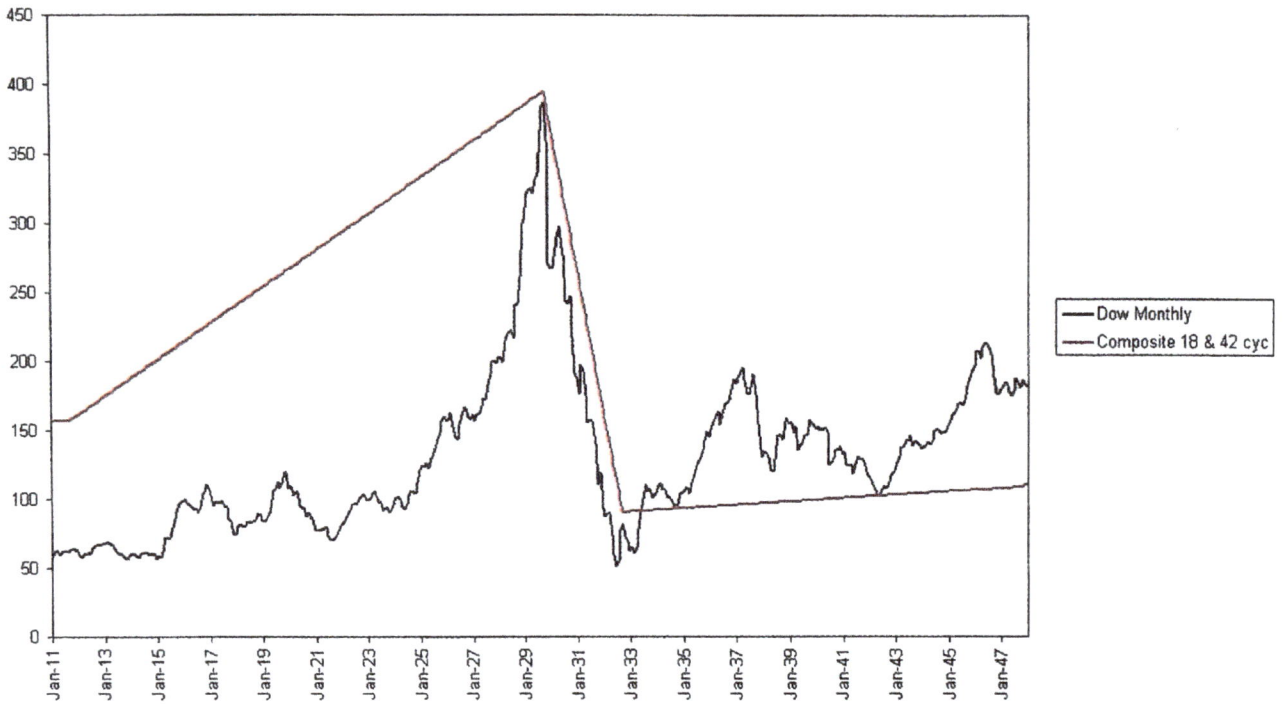

Composite 1947-1982 ©

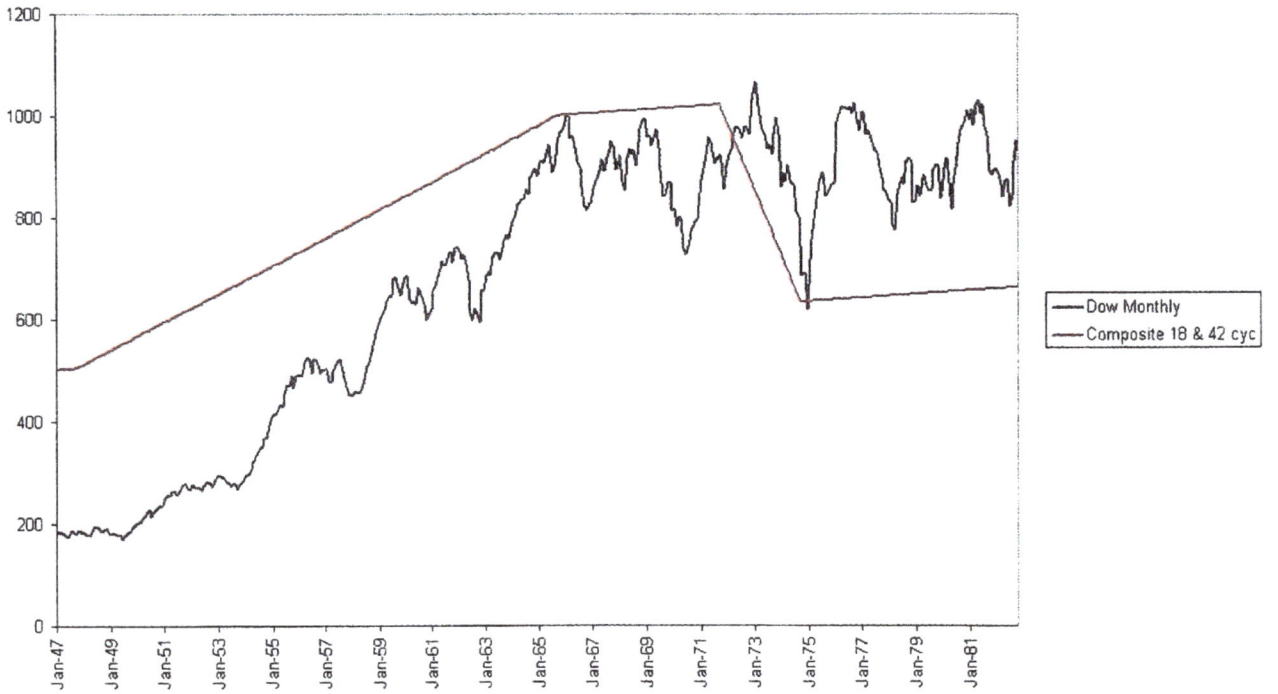

134

can't guarantee that these two cycles, which have been present in the data for several decades, will continue; but I for one would not bet my own money against it! If we project this simple model of market behavior into the future, we get the following results.

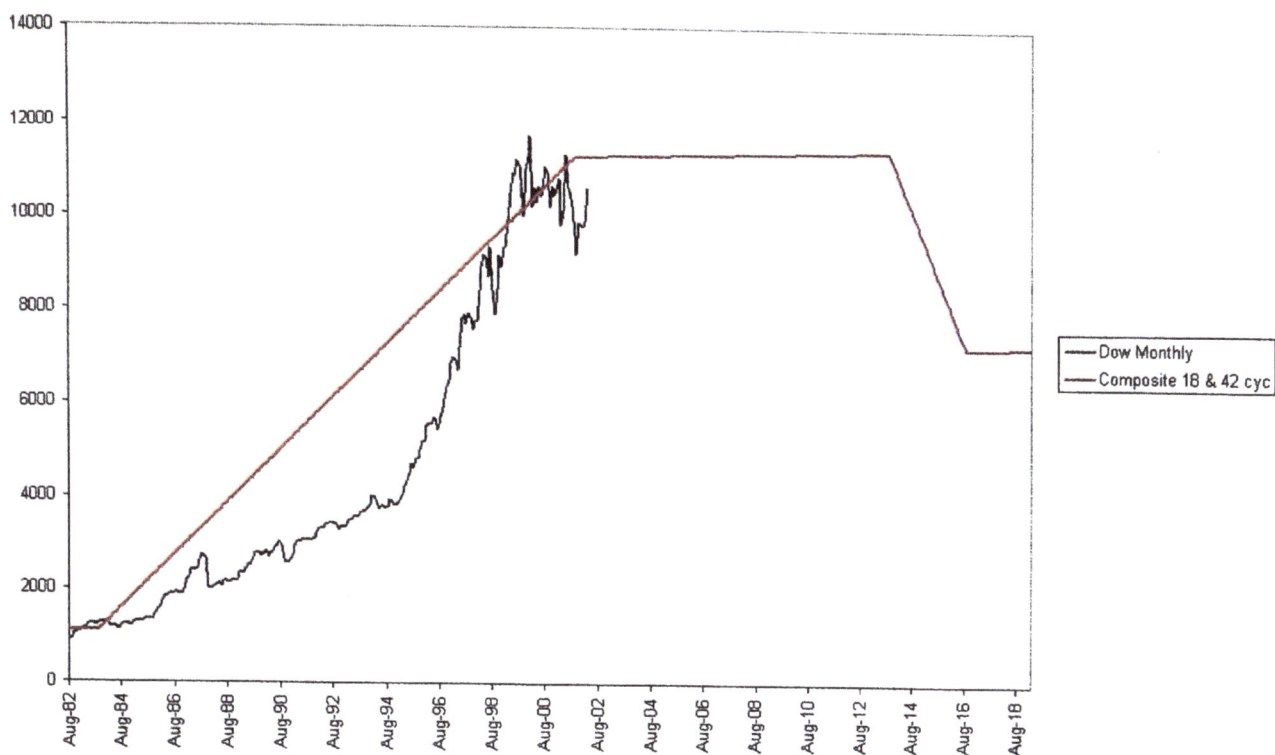

or those that are confused as to how this model was created, it is simple addition. I took the 42-year pattern and added it to the 18-year pattern. The flat areas are the result of one cycle that is still going up while another cycle is going down, which causes them to cancel each other out. The really strong advances and declines are the result of both cycles moving in the same direction.

his cyclic model suggests that the equity markets are currently in a basic flat consolidation period with a very mild upward bias due to the 39-year advancing phase of the 42-year cycle until approximately the year 2013. Please understand that this projection is **not** saying that the Dow Jones Industrial Average will reach 7200. It could go a lot lower or it could go higher. This simply projects the basic shape or trend of the pattern. There is likely to be some kind of "financial panic" after this point (2013) that takes the market down to make its 42-year cycle low in the year 2016. Remember that this has historically been an extremely important stock market low. There is an "appointed" time to reap and a time to sow. This applies to long term investing as well. In Genesis chapter 41, Joseph interprets the dream of the Pharaoh. Joseph tells the Pharaoh that his dream simply means that there will be 7 years of abundance followed by years of famine. Because Joseph was able to interpret the Pharaoh's dream, he was put in charge of the whole land of Egypt. During the 7 years of plenty,

Joseph had stored the entire surplus from the crops to be used when the famine came. ***"When the 7 years of abundance enjoyed by the land of Egypt came to an end, the 7 years of famine set in, just as Joseph had predicted. Although there was famine all over the rest of the world, food was available throughout the land of Egypt."*** Knowledge of this 7-year cycle did not prevent the famine from coming at its appointed time. The knowledge simply allowed Joseph to prepare for the situation. **Forewarned is forearmed.** There are other cycles that we could add to our model to get a more accurate representation of market behavior, but for the majority of investors, this is really not necessary. *"My experience has taught me that nothing can stop the trend as long as the time cycle shows an up-trend. Nothing can stop its decline as long as the time cycle shows down"* —W.D. Gann, *45-Years in Wall Street,* page 3. I will point out that there also appears to be a strong 8-year cycle in the data. For example, there are some very pronounced lows every 8 years (1942, 1950, 1958, 1966, 1974, 1982, 1990, 1998). Next one would be 2006. Since the 42-year cycle is still up trending until 2013, this could potentially be a nice place to invest in aggressive growth mutual funds for the next six to seven years. In a "bear market," you are typically better off in "growth and income" style mutual funds or "balanced funds." These funds typically invest in stocks that pay dividends, which usually perform better than non-dividend paying stocks when the equity markets get weak. Also, they typically own some fixed income investments, such as bonds and preferred stocks, which help the fund weather out the storm in "bear markets." As a simple rule, it's typically better to be in "aggressive growth funds" when the 18-year cycle indicates a "super bull market" and when a "super bear market" is indicated; you may want to be more conservative in your investment approach. Using a "growth and income fund," a "balanced fund" or some other more conservative investment vehicle during these tough times may not be a bad choice if you want consistent exposure to the equity markets. For those interested in a possible astronomical correlation to the 42-year cycle, I suggest that you investigate heliocentric Uranus. For those of you that want to continue to project this composite cycle further out into the future, you can do so with simple addition. Again flat periods are caused when the cycles cancel each other out, i.e., one is going up and the other is going down and strong trends are the result of both cycles going in the same direction. Therefore, after the 42-year cycle low in 2016, the markets should advance in a sideways pattern similar to the years 1932 to 1947 and 1974 to 1982. However, this sideways market will only last approximately two years because the 18-year cycle should bottom late in the year 2018, starting the next "super bull market," which should be strong until 2036. From 2036 to 2055, the markets should be stuck in another sideways pattern like the one in 1932 to 1947 followed by a "panicky break" down into the year 2058. Because the 18-year cycle bottoms in 2054, the 42-year low may not come down as hard as it has historically, i.e., it will have less momentum than usual. In any case, based on these two cycles, the market should be extremely strong from 2058 to 2072. I have included a chart at the end of this report that projects the 18-year and 42-year cycle composite out a full 100 years to the year 2102!

INTEREST RATES- I should also point out something about interest rates. Many "financial experts" will claim that declining interest rates create a strong economy and thus lead to a strong bull market, while high interest rates tend to slow the economy and thus lead to weaker markets or even bear markets. As a general rule, this seems to be true, but I need to point out that the 1929 crash and Great Depression took place when the Federal Reserve was aggressively lowering interest rates.

Japan's equivalent stock market depression, which has run from about 1990 to 2002 (so far), has also occurred in a declining interest rate environment. The same can be said for the current stock market sell off in the United States. The NASDAQ declined approximately 80% in a declining interest rate environment and the S&P500 declined nearly 40%. Therefore, if the past is any guide, the worst "bear markets" in history have happened in a declining interest rate scenario. Maybe it's because the government is trying to fix a problem that has already gotten way out of hand?

THE ECONOMY- In general, the economy usually runs in cycles of 4.65 years or 4 years and 8 months if you prefer. As a basic rule, the economy will advance above normal for 4.65 years, then decline back to levels that are more normal over the next 4.65, then drop to below normal over the next 4.65 years and then return back to normal, where the cycle starts all over again. This cycle in total measures about 18.6 years, which is very similar to the 18-year super bull and bear market cycle just discussed. Under this basic scenario, the economy was due to peak in its above normal phase in August of 1999. It will continue to decline for 9.3 years (9 years and 4 months) where it will reach its below normal level. This basically means that economic conditions will slowly decay until the year 2008, when conditions will start to improve again. This cycle is harder to find because the "wealth affect" that is created in the 18-year super bull markets tends to hide this cycle. But if you subtract 4-years and 8-months from August 1999, you get the year 1994, which was considered to be a mini-recession during Clinton's first term as president. In any event, I have found this to be a fairly useful tool since economists, as a general rule, cannot seem to do anything more than analyze the past. Based on this cycle, I was prepared for the new economy to reach its peak in late 1999. Again, forewarned is forearmed. This particular cycle was based on the work of Louise McWhirter and the North Node of the Moon. In 1875 Samuel Benner wrote a small book called *Benner's Prophecies of Future Ups and Downs in Prices*. Benner's pig iron price forecast, which came true over a 60-year period, produced a gain to loss ratio of 45 to 1. Benner also created a forecast chart of financial panics titled **Periods in Which to Make Money,** which was probably made during the Civil War. In this chart, Benner forecasted a high in business activity and thus a high for our financial markets for the year 1972 followed by a panic type low for 1981, a very good call. The actual low came in 1982, but a good call nevertheless. Benner then called for a business activity top in 1989 as he was anticipating a major recession between 1994 and 1996, which was another very good call as 1994 certainly experienced a recession. He expected 1996 to be a great buying opportunity with his next panic due in 1999. Considering this forecast was made in the 1800's, and the actual panic started one year later in 2000, I feel that this was yet another great call. I bet there are plenty of people that wish they would have sold in 1999! Benner's forecasting technique was unique because he did not see cycles as symmetrical sine waves. Instead, Benner separated his cycles into number sequence patterns that were broken up into very simple rhythms. Since this section of the report deals with the economy, I will give you Samuel Benner's forecasting pattern for "business activity," Periods of "good times and high prices," i.e., the time to sell stocks and other values of all kinds. Their cycles are **8-9-10** years and repeat. For example, 1972 was the last projected "business activity" top in the sequence. Adding 8 years projected 1980 as the next "business activity" top. Next, adding 9 years to 1980 projected 1989 as a "business activity" top. Adding 10 years projected 1999 as a "business activity" top. Now the cycle starts over again and we add 8 years to project 2007 as a "business activity" top. Periods of "hard times and low prices," a good time to buy stocks and other values and hold them until the next "business activity" top and then unload are based on the sequence **7-11-9** years and repeat. The last low in this sequence was 1978. Adding the **7-11-9** year sequence gives 1985, 1996 and 2005 as projected times to buy based on Benner's work. "Years in which panics have occurred and will occur again," are projected with the sequence **16-18-20** years and repeat. Obviously, you would ideally want to unload stocks, mutual funds or other values prior to a panic and buy them back at lower prices after a panic has occurred. 1965 was the last projected panic in this sequence. Adding the **16-18-20** year pattern to this date gives 1981, 1999 and 2019 as Benner's "years in which panics have occurred and will occur again." Based on these three sequences (good times, hard times, and panics) one would have looked for a buying opportunity around the "panic" year 1981 to hold into the projected "boom" year 1989. The next buying opportunity would be the "hard time" year 1996, which would be held until the "business activity" top in 1999 where one would start unloading stocks prior to the projected panic. Please study these simple number patterns and examples as Samuel Benner's work has certainly stood the test of time! A reprint of Benner's book can be ordered from the Sacred Science Institute (800) 756-6141 or www.SacredScience.com.

THE JANUARY EFFECT- I have heard many of the "experts" make mention of the January effect. There is a historical tendency for the market to rise in the month of January. I even remember reading an article in the Wall Street Journal or Investors Business Daily that talked about some guy who had made a fortune selling put options on the last trading day in December so he could buy them back cheaper in January and make a profit. People who buy put options are betting the market is going down. He was selling them, so he was expecting the markets to go up. If he was right, the options would become worthless and he would keep all the money he received from selling them. Anyway, this person made something like $300,000 doing this over the course of 1994 to 1998 and reinvesting all his profits each year. I know for a fact that if he kept repeating this same strategy, he had to financially destroy himself in the years 2000, 2001 and 2002. Now, January does have some predictive value for those that know how to read the signs. Basically, if the last week in January closes lower than the first week, there is a tendency that the whole year is likely to finish lower. If the last week closes higher than the first week, there is a tendency that the whole year will finish higher. Examples: The last week of January 2000 closed lower than the first week and the market finished that year lower. The last week of January 2001 closed lower than the first week and again the market ended that year lower. The last week of January 2002 closed lower than the first week and it remains to be seen if this particular year will end lower. This is just a very basic tool that can be used each year to get a feel for what the market might do for the remaining 11 months.

I have made every attempt to make this information affordable so that it reaches those that need it most. Over the last 10 years, I have personally spent well over $30,000 in obtaining knowledge about the financial markets. Most of the information was not worth the price I paid for it. This cycle work has been an extremely valuable roadmap for my long term investment decisions, which in my opinion makes this small little report worth much more than the small sum of money it took you to obtain it! After all, you have a basic market forecast that is good for the next 100 years or more!

Please refer this report to those you care about. Do not make them a photocopy, but let them purchase it for themselves, as it is very affordable for anyone. As I have already said, I have spent thousands upon thousands of dollars of my own money learning what works in the markets. By making this report available to you at a fair price, I have been very generous with my knowledge and discoveries. After all, it did not cost you over $30,000 and a solid 20 years of your life to obtain it! So please do not just give it away for free! People tend to have little respect for information they obtained for nothing. The same rule applies to the research work of others! Hold yourself to a high standard of integrity and you will automatically gain the respect of others. I sincerely wish you great success in the near future concerning your life, happiness, personal freedom and of course your financial investments.

Sincerely Yours,

Daniel T. Ferrera

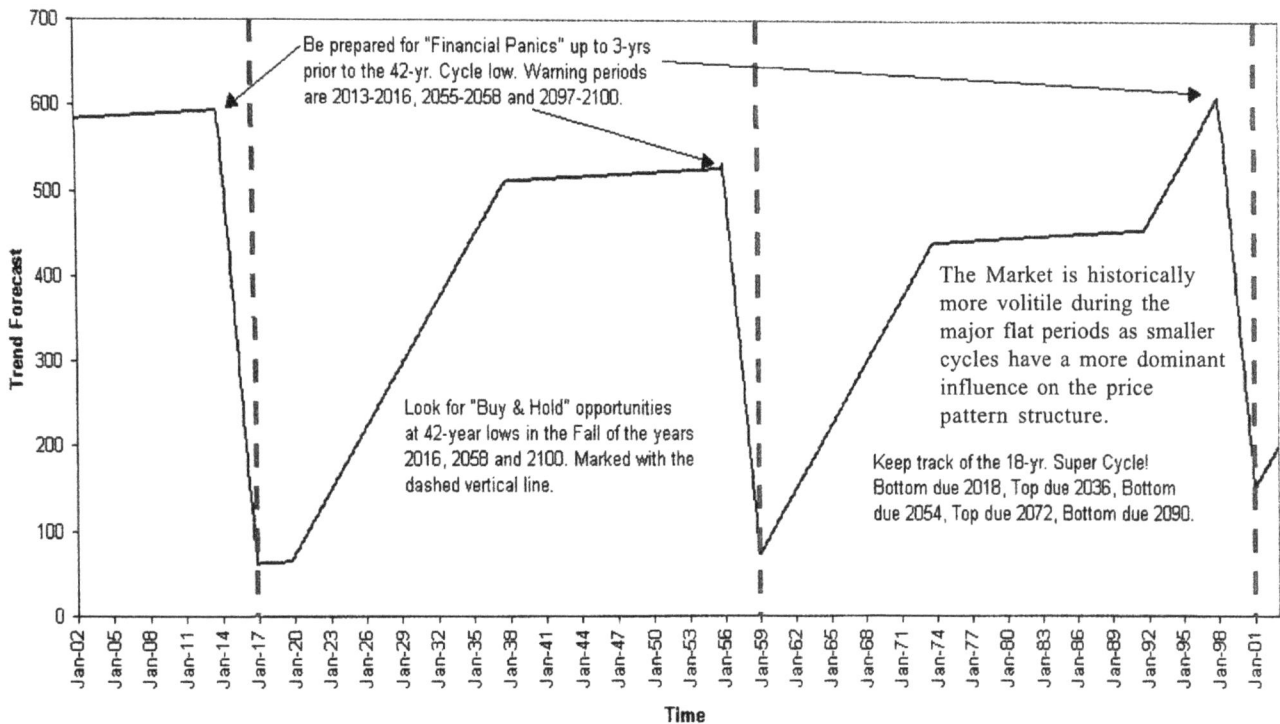

THE SECRETS OF FORECASTING II

In the October 2002 "20th Anniversary Issue" of <u>Technical Analysis of Stocks & Commodities</u>, there was a small section dedicated to former market gurus called "The Titans of Technical Analysis." In particular, page 34 of this issue had a very short piece on W.D. Gann, which I would like to discuss further. W.D. Gann was famous for his ability to forecast markets. He would map out and draw a curve or projection of the stock market, cotton, grains and a variety of other markets one full year in advance. Gann stated that this forecasting ability was based on his "Master Time Factor" and "Cycle Theory." Gann's basic belief was that since supreme universal laws operate on Earth even down to the tiniest atom, which is commonly accepted, all actions of man or of the products of the earth are also guided and directed by these same universal laws. ***"By the Law of Periodical Repetition, everything which has happened once must happen again and again and again-and not capriciously, but at regular periods, and each thing in its own period, not another and each obeying its own law...the same Nature which delights in periodical repetition in the skies is the Nature which orders the affairs of the earth. Let us not underrate the value of that hint."—Mark Twain.*** In short, "Like causes will produce like effects."

Today, there is a great multitude of techniques used in technical analysis to anticipate and/or forecast possible future movements of the markets. Some techniques, such as the "Elliott Wave," are based on a type of pattern recognition. Under this method of analysis, trends basically consist of 5 impulse waves in a main direction followed by 3 corrective waves against the main trend. Price swings are viewed as being part of smaller and larger structures completing a type of growth spiral in a somewhat predictable sequence. Levels of support and resistance are projected using Fibonacci ratios of prior impulse waves. More traditional forms of technical analysis pattern recognition include the classic chart formations such as the "head & shoulders," "double tops & bottoms," "triangles," "cup & handle," "bull & bear blags," and "Japanese candle sticks," to name a few. Other techniques rely on the measurement of market volume and/or open interest either individually or in combination with market patterns. Then there is the use of technical indicators such as moving averages, relative strength, stochastics, momentum, etc., which are used to form opinions about the current condition of the market and thus anticipate some kind of future direction. Of all the techniques available, it is the use and knowledge of time cycles that allows one to project and/or potentially model the future price behavior! This is possible because price action seems to unfold in repetitive rhythms over time. It is my belief that W.D. Gann used his knowledge of time cycles to make many of his incredibly accurate market forecasts in his day. Gann believed that " Time is the most important factor in determining market movements" because everything moves in cycles, as the result of the natural law of action and reaction. "By a study of the past, I have discovered what cycles repeat in the future." Gann was not the only person in his day to use cycles to amass significant wealth. The extremely rich Rothchild family along with individuals like J.P. Morgan also used cycles to manage their investment decisions as well.

In 1875, Samuel Benner wrote a small book called *Benner 's Prophecies of Future Ups and Downs in Prices*. Benner's Pig Iron price forecast, which came true over a 60-year period, produced a gain to loss ratio of 45 to 1. Benner also created a forecast chart of financial panics that was titled "Periods in Which to Make Money," which was probably made during the Civil War. In this chart, Benner forecasted a high in business activity and thus a high for our financial markets for the year 1972 followed by a panic type low for 1981, a very good call. The actual low came in 1982, but a good call never the less. Benner then called for a business activity top in 1989 as he was anticipating a major recession between 1994-1996, which was another very good call as 1994 certainly experienced a recession. He expected 1996 to be a great buying opportunity with his next panic due in the year 1999. Considering that this forecast was made in the 1800's, and the actual panic started only one year later in 2000, I feel that this was yet another great call. I bet there are plenty of people that wished they sold their stock portfolios in 1999! Benner did not see cycles as symmetrical sine waves. His interpretation was based entirely on number sequences, which I believe influenced W.D. Gann. On page 75 of *Tunnel Thru the Air,* Gann wrote that his "calculations are based on the cycle theory and on mathematical sequences. History repeats itself." Anyone who reads Benner's *Prophecies ofFuture Ups and Downs in Prices* will certainly find many similarities with several of Gann's published works.

In 1944, Edward R. Dewey, President of the Foundation for the Study of Cycles, produced a ten-year stock market forecast based on a synthesis of 10 different cycles that produced a gain to loss ratio of 185 to 1. The forecast and corresponding chart can be found in his book *Cycles, the Mysterious Forces That Trigger Events.* This was a long-term projection based on 10 large cycles ranging from 4.5 to 27-years in length.

In another quote from *Tunnel Thru the Air,* Gann said "Mathematical science, which is the only real science that the entire civilized world has agreed upon, furnishes unmistakable proof of history repeating itself and shows that the cycle theory, or harmonic analysis, is the only thing that we can rely upon to ascertain the future." The Foundation for the Study of Cycles defines "harmonic analysis" as a method of describing a time series (like price and time). It is accomplished by fitting sine-cosine curves of harmonic (unit fraction) lengths to the data series. When combined, the sine-cosine curves so obtained will reproduce the original data series curve to any degree of accuracy desired, depending upon the number of sine-cosine curves used. Based on his approach, W.D. Gann claimed that it was just as easy to forecast the upcoming year, as it was to forecast the next 100 years based on his cycle theory. After many years of research, I can certainly understand his opinion. Along with this article, I'm including a chart I created that attempts to model the markets movements over the last 112 years using only the summation of 16 different cycles and amplitudes. Overall, this chart has been a very good model of the stock market and it can easily be projected out for the next 100 years just as Gann claimed. "The future is just a repetition of the past." I have recently finished a book on market cycles called *Wheels within Wheels, The Art of Forecasting Financial Market Cycles,* which will include the 100 year stock market projection based on these same 16 cycles. Examples of shorter term forecasting based on cycles of shorter length are also provided in the material. All of my personal Excel worksheets and stock market data will be included with the material so that you can duplicate the basic process for other markets. I have already received many nice e-mails telling me that the 100 year projection alone is worth several times the price of the entire book.

The wheels material covers both the long-term forecasting and more intermediate term forecasting of the U.S. Stock Market using purely cyclic models. In addition to the stock market cycles, I have also provided some long-term cycle work for the bond market (interest rates) and gold. Recently, as a favor to a student, I added a cycle spreadsheet and projection for the Australian Dollar as a free bonus. Contact Traders World, The Sacred Science Institute (800)756-6141 or www.SacredScience.com for availability. For those that would like an alternative to ordering the "Wheels Within Wheels" material, I have a "Special Stock Market Report" that illustrates two dominant long term cycles in the stock market and projects this pattern 100-years as a simple model. This report can be ordered at www. tradersworld.com for $29.95 with a 100% money back guarantee. Serious cycle students should also obtain a copy of Edward Dewey's course *How to Make a Cycle Analysis* from The Sacred Science Institute. This 630-page course was written in 1955 and has somehow recently resurfaced. I have just obtained a copy and must say that it is simply the best work on cycle analysis that I have ever seen. I certainly wish I had obtained this information several years ago!

Daniel T. Ferrera

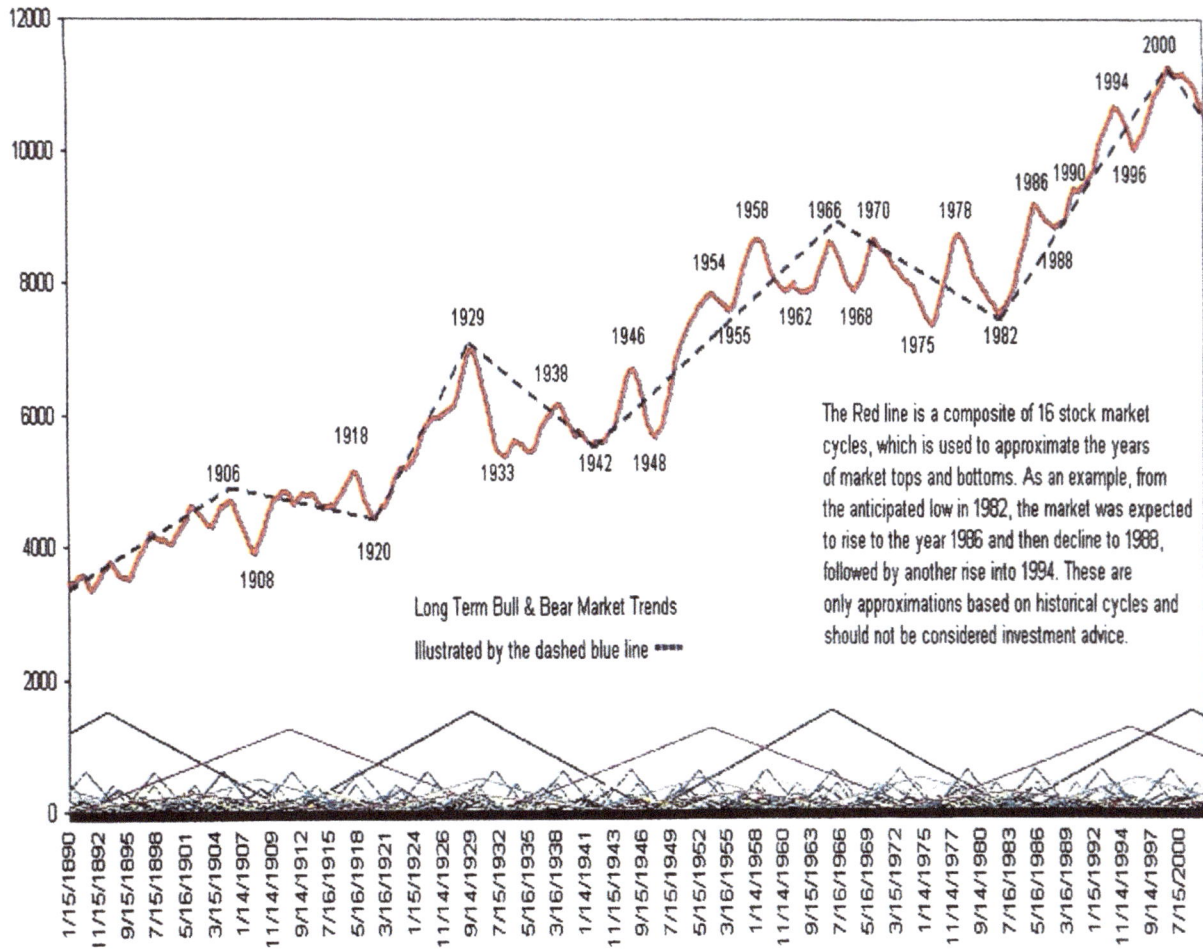

DTF Long Term Stock Market Barometer a Composite of 16 Cycles

The Red line is a composite of 16 stock market cycles, which is used to approximate the years of market tops and bottoms. As an example, from the anticipated low in 1982, the market was expected to rise to the year 1986 and then decline to 1988, followed by another rise into 1994. These are only approximations based on historical cycles and should not be considered investment advice.

Long Term Bull & Bear Market Trends
Illustrated by the dashed blue line ----

16-Cycle Baromter 1982-2003

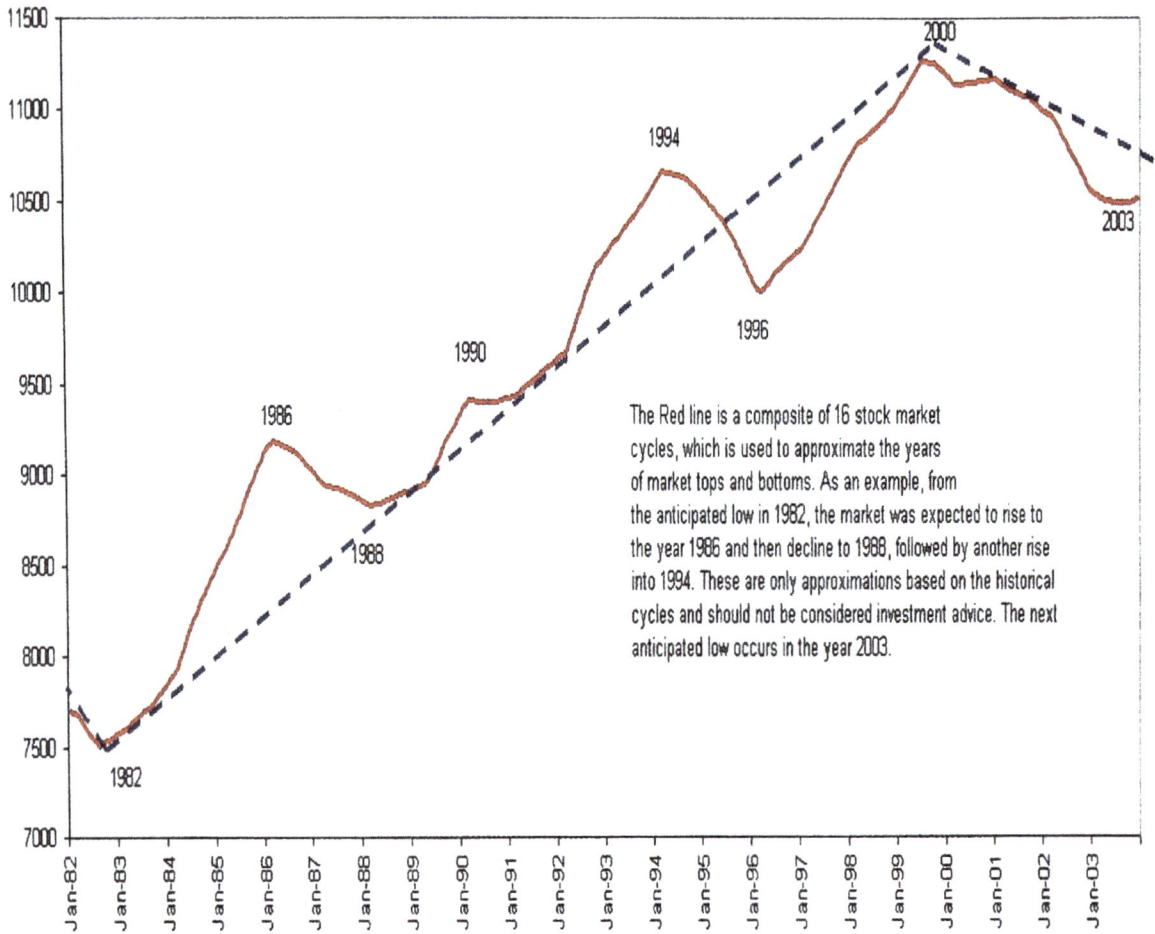

The Red line is a composite of 16 stock market cycles, which is used to approximate the years of market tops and bottoms. As an example, from the anticipated low in 1982, the market was expected to rise to the year 1986 and then decline to 1988, followed by another rise into 1994. These are only approximations based on the historical cycles and should not be considered investment advice. The next anticipated low occurs in the year 2003.

W.D. GANN'S ANNUAL FORECASTS

During his career, W.D. Gann made annual forecasts available to the public for a variety of markets, including the stock market averages, grains and cotton. These forecasts sold for $100 during the Great Depression in the 1930's, which would be equal to paying over a $1000 in today's money. Gann always kept the basis of this forecasting technique veiled in secrecy, primarily due to the simplicity of the method. After all, if the general public knew how these forecasts were produced, they probably would not have been willing to pay such high prices for them. The basis of the method is, in my opinion, almost exactly the technique described in Professor Weston's book, *Forecasting the New York Stock Market,* also known as *The Geometrical System of Forecasting.* If you make a study of Gann's Forecasting Course and seriously compare it to Weston's, I think you will find them amazingly similar. For example, both Gann and Weston believed that the prior 10-year and 20-year intervals or cycles were the key to anticipating what the market would do in the present year. Although Weston used monthly data, and Gann used daily data, you will find that their respective forecasts for 1929 are basically identical in terms of the general trend of the market. I have stated before that the future is but a repetition of the past; therefore, to make up a forecast of the future, you must refer to the previous cycles. Gann wrote in his "Master Forecasting Course" that: "The previous 10 year cycle and the 20 year cycle have the most effect in the future, but in completing a forecast, it is best to have 30 years past record to checkup, as important changes occur at the end of 30 year cycles." Professor Weston illustrated the same thing and specifically stated that he believed the phenomenon was caused by the angular distances between the planets Jupiter and Saturn. When you compare Professor Weston's 1929 forecast with Gann's, you will see that Professor Weston simply averaged out the years 1878, 1889, 1898, 1909, and 1919. Gann utilized the years 1869, 1909, 1919. The main difference in the two was that Weston used the years 1878 and 1898. Professor Weston explained in his work that due to the varying speeds of Jupiter, Saturn and the apparent retrograde motion as viewed from the Earth, at certain periods, some calendar years would repeat as both the ending of a prior cycle and the start of a new cycle while other years are skipped over entirely. Now, since Gann was very much into numerology as well as astrology, he most likely eliminated the years ending in "8" to make his forecast. The basic approach that underlies both Gann and Weston is to locate and line up prior years with the angular relationship of Jupiter and Saturn. Once you have found a few of these "analogous" years, you simply average them all together to create a forecast for the current year or cycle. That is the basis of their entire forecasting method in a nutshell. This "geometrical forecasting system" can be applied to shorter time periods as well and the level of detail can be taken to many degrees of intricacy once you understand the basics. As an example, one of my course owners decided to apply Gann's and Weston's use of Venus in predicting the stock market. Weston comes right out and states that it is Venus, while Gann simply calls it a 7-month cycle. In any event, this "student" simply took the average of the prior five cycles of heliocentric Venus and used the data to create a projection for the current cycle. In other words, he simply took five complete 7-month cycles in the S&P500 and averaged them all together to create an average projection of what the next 7-month cycle should basically do.

As you can see from a study of the chart below, the market is basically following the projected stock market curve made from averaging the prior 5 cycles of Venus. This is the same technique that Professor Weston illustrates in his book and is in my opinion directly related to Gann's overall forecasting approach. Adding the Mercury cycle would further enhance the forecasting resolution. In closing, I suggest that you seriously study the basics of the longer-term forecast, then refine and fine-tune the approach by doing the exact same thing with the smaller cycles that are contained within the larger cycles. This is the main reason why Gann created charts like "The Master 20-Year Forecasting Chart."

Helio Venus Pattern - SPX

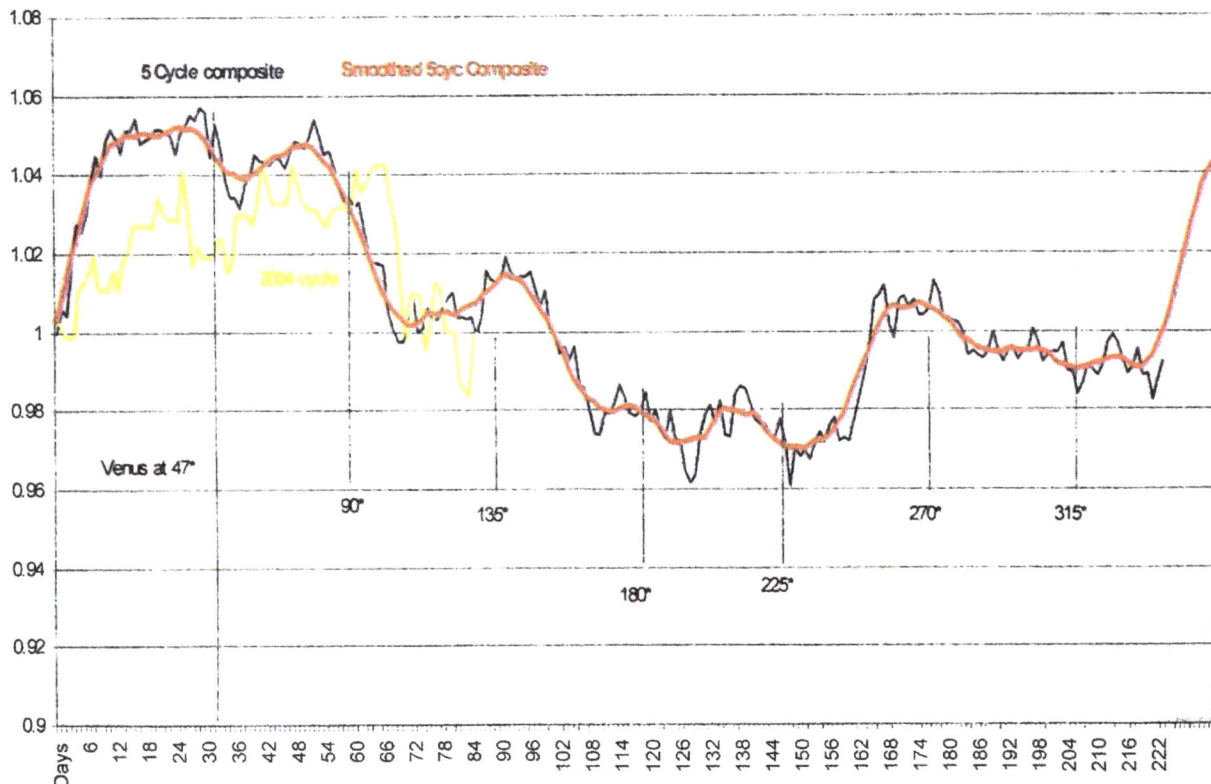

He wanted a simple method of stacking each 20-year period of market data on top of one another so he could see when tops and bottoms were likely to occur in the present cycle. There is no other reason for making charts of this type. In today's world of cycle analysis, this method is more commonly known as array analysis. For all the mystery, hype and high cost associated with these forecasts and "The Master Time Factor," it certainly appears that it was a relatively simple concept. Another market prediction technique, that is really just another array analysis, is the Delta Phenomenon by Welles Wilder. This "Phenomenon" is primarily related to the sun and moon. Once you find a cycle time period that consistently turns the market, array analysis will give you the cycle's basic shape.

THE KEYS TO SUCCESSFUL SPECULATION
Daniel T. Ferrera

2004. 473p. + supplements. The Keys To Successful Speculation presents the first fully intact trading manual, which will clearly teach anyone, including those with absolutely no prior knowledge of the markets, Gann, or forecasting in general, to successfully trade in any market, from stocks, to futures, to options, in any time frame, from day trading to long term trading, beginning with very limited capital, as little as a few thousand dollars. This book is not esoteric, and is not focused on abstract analytical and theoretical principles, but is totally application orientated, specifically outlining a clear trading strategy incorporating all necessary principles of money management, charting, risk management, swing trading, signal generation, the use of options and much more. Contents: Introduction; Dedication; Risk Discloser; Commodity Basic; How Much Do You Need To Start Correctly; The Profession & Business; Charts; The Keys To Successful Speculation; Mathematical Analysis Section 1; Perspective; Mathematical Analysis II Geometry Review; Instantaneous Balance Stability; Gann' Mechanical System; Swing Trading Improvements Theory & Principle; Understanding and Exploiting Lost Motion; Endpoints of Swings Have Magnetic Force; Zones of Influence Define Test & Failure; Simple Profit Targets & Market Examples; Trading Pattern #2; How Price Can Change Polarity; The Fourth Time at The Same Level; How to Determine A Useful Swing Size; How To Determine Useful Stops; Trading Is A Profession; Understanding The Options Opportunity; Probability of A Price Move; Understanding Option Spreads; Successful Trading ; What Is Luck; Conclusion; Opportunist System; Swing Trading Stocks Based on Square Roots; Questions & Answers; Understanding Trends & Trendline Breaks; Trends Again, Bar Grouping Technique; Gann's Red Light Green Light Trend Indicator; Using Inside Bars to Enter With the Trend; Gann's Natural Resistance Levels & Cycle Points. Includes CD ROM with TradeStation Indicators.
Deluxe Quarto Hardcover Edition. Black Suede w/Gilt Lettering. CAT#500 $1500.00

THE PATH OF LEAST RESISTANCE
The Underlying Wisdom & Philosophy of W. D. Gann Elegantly Encoded in the Master Charts.
By Daniel T. Ferrera

2015. Text: 298 Pages, Appendices: 100 Pages. Numerous Diagrams. Software Included. The main purpose and objective of this course material is to provide the reader with a comprehensive yet practical understanding of W.D. Gann's most useful trading tools and their many uses. Over his 50 year trading and advising career, W.D. Gann developed approximately 40 different trading tools, calculators and/or mechanical methods to trade with and many of them are presented in this very detailed course. The material in this book will explore Gann's different techniques and tools for trading according to his rules, and is intended to be a very practical guide to be used in conjunction with the risk management principles and account management rules presented in my earlier course, The Keys to Successful Speculation. CONTENTS: The Basis Of Gann's Method; Important Clues, 4th Dimension; Base 10 Method; Parallel Angles; Squaring Or Balancing Price Charts; Base 10 Fractal System; Base 10 summarized; Visual Balancing Method; Squaring Price With Time; Three Ways To Square Time And Price; The Most Important Angle, Diagonal; Second Most Important Angle, Horizontal; Third Most Important Angle, Vertical; Sections Of Bull And Bear Markets; Short Summary; Simple Time Periods; Balancing Time With Equal Time; Summary Of Simple Time Periods; Short Term Time Projections; Fibonacci And Natural Squares; Formations & Form Reading; Swing Indicator; Where To Place Stop Loss Orders; Formations Review; The Halfway Point; Moving Average Trend Indicator; The Law Of Vibration; Price Changes Are Vibrations; Let The Market Tell Its Own Story; Support And Resistance Levels; Odd And Even Squares; Fibonacci Price Projection; Square Roots; Price Projection Method; Projection Process Applied; Money Management; The Square of Odd & Even Numbers; The Spiral & Square Charts; Time Counts & Quick Calculations of Angles; Conclusion. Appendices: All Gann Courses & Materials Referred to...
CAT#300 Deluxe Quarto Hardcover Edition. Black Suede w/Gilt Lettering. $1995.00

ECONOMIC & STOCK MARKET FORECASTING
W. D. Gann's Science of Cyclical Periodicity Sequencing
Daniel T. Ferrera

2013, 285p. Numerous Diagrams. Software & Data Included. After seven long years of waiting since Dan Ferrera wrote his last course The Spirals of Growth & Decay, we are very happy to announce the release

of his newest and latest work, Economic & Stock Market Forecasting, W. D. Gann's Science of Cyclical Periodicity Sequencing! During these last years, while Ferrera's interests have focused elsewhere, he has continued to write his Yearly Outlooks, giving a forecast for the economic environment and general stock market. These Outlooks have continued to be very popular, and have honed Dan's analysis and forecasting abilities to a new level, leading him to a NEW BREAKTHROUGH in his understanding of Gann's most complicated and secretive forecasting methodology, what we would call Cyclical Periodicity Sequencing. The intent of this course is to explain Gann's science of Mathematical Sequencing of periodic market patterns, and how it works in conjunction with Gann's "cycle theory", in order to forecast the general economy, stock market, or individual stocks, through identifying the periodic sequences of market action. It also presents a study of Key Options Strategies and techniques to take advantage of these forecasts for both short and long term trading. The course presents new material upon one of Gann's very deepest levels of analysis, which to our knowledge has never been so clearly and applicably presented in any other work prior to this. CD ROM Including Excel Cycle Modeler, Planetary Cycle Calculator & Dow Jones Data 1790-2010 Monthly, 1900-2013 Daily. INCLUDES FERRERA'S OUTLOOKS FROM 2009-2014! MEMBERSHIP IN ONLINE FORUM INCLUDED!
CAT#444 Deluxe Quarto Hardcover Edition & CD Rom. Black Suede w/Gilt Lettering. $2995.00

MYSTERIES OF GANN ANALYSIS UNVEILED!
Daniel T. Ferrera
2001, 350p. This course presents the most detailed explanation of Gann Theory &Application ever before released to the public. Covering in detail a dozen of Gann's most difficult analysis techniques and theories, this course will advance the general reader to levels far beyond most Gann "experts" known today. Contents: The Astrological Secret Of Gann Angles; Analysis Of The Coffee Letter & Planetary Vectors Or Angles; The Cosmology Of 17 Years; Squaring Price With Time; Forecasting With Planetary Cycles; The 37 Year Cycle Pattern In The Dow Jones Industrial Average; Important Formulas And Techniques For Planetary Cycles; Three Term Proportion Of Planetary Longitudes; Gann's CE Average, MOF Formula & Master Charts; Periodic Number Cycles; Support & Resistance Techniques From The Square Of 9; Converting Planetary Longitude To Price; Converting A Horoscope Into A Price Calculator; Gann's Master Mathematical Formula For Market Predictions; Mass Pressure Forecasting Technique; Forecasting The Stock Market With Cycles; Gann's Permanent Charts; Gann's Base Ten Method; Balancing Solar Longitude With Price On The Square Of 9; Tunnel Thru The Air; Soybean Letter To Private Student; Market Volume; Conclusion. Includes 6 computer programs to calculate the various Gann techniques listed above.
CAT#439 Deluxe Quarto Hardcover Edition. Black Suede w/Gilt Lettering. $1,500.00

WHEELS WITHIN WHEELS:
The Art of Forecasting Financial Market Cycles.
Daniel T. Ferrera
2002, 230p. Numerous Diagrams. The newest course by Dan Ferrera, breaking down the 16 primary component cycles of the DOW Jones Averages, producing an accurate map of the last 100 years of history, and projecting the cycles ahead to 2108. Includes all Excel Spreadsheets with all cycle calculations and charts, and the 100 year projection DFT Barometer. Contents: PART I - Special Stock Market Cycle Report; Author's Introduction; The 18-Year Super Bull & Bear Market Cycle; The Big Picture; A Closer Look At Cycles; The 42-Year Cycle; Interest Rates; The Economy; The January Effect; 2002- 2102 Major Trend Cycle Composite Forecast; PART II – Special Stock Market Cycle Report; Author's Introduction; What is a Cycle?; The New Era; The Four Primary Intermediate Cycles; W.D. Gann's Stock Market Patterns; The 10 & 9 Year Cycles; The Shorter Cycles; Putting Them All Together ; S&P 100 Year Projection Using #1 & #2 Dominant Cycles; Follow The Yellow Brick Road; PART III – The DFT Long Term Stock Market Barometer; 16 Cycle Composite Barometer; Is Timing The Market Worth The Effort?; The 54-Year & 12-Year Cycles In Bond Yields; Cycles In Gold; Stock Market Cycle Charts; APPENDICIES: 1 - Garrett Torque Analysis Example; 2 – How To Create A Composite Cycle; 3 – Vectors & Phase: What is a Vector?; 4 – Understanding Cycles; 5 – Wyler's Theoretical Considerations; 6 – Dewey's Cycles In The Stock Market; 7 – Cogan's Rhythmic Cycles; 8 - Chase's Economic Time; 9 – Wood's Stock Market Time Cycles; 10 – Martin's Trend Action; 11 – Weston's Geometrical Chart System;

12 – Bibliography & Recommended Reading. CD ROM Including Excel Cycle & DFT Worksheets.
CAT#447 Deluxe Quarto Hardcover Edition. Black Suede w/Gilt Lettering. $450.00

THE GANN PYRAMID: *SQUARE OF NINE ESSENTIALS*
Daniel T. Ferrera

2001. A new groundbreaking course on the Square Of Nine, W. D. Gann's most mysterious calculator. This course is full of never before seen principles and techniques of analysis using Gann's Square of 9, with detailed explanations of their applications to the markets. Introduction; Navigating With the Square of Nine; Bible Interpretations Related to W. D. Gann; What Gann Said About the Square of Nine; Six Squares of Nine; Square of Nine Time Applications; Price Targets For Support & Resistance; Using A Square of Nine Table; Time As a Price & Price as a Time; Gann Angle Projection; Square of Nine Time Techniques, A Different Look at History; Analyzing Markets; Nine Rules For The Square of Nine; Periodic Number Cycles; Price as a Time Period; Price Levels For Support & Resistance; Converting Astronomical Longitude to Price; Another Astronomical Technique; Fibonacci Ratios; Conclusion; W. D. Gann Calculators. Includes an Excel Square of Nine Calculator, and Plastic Overlay.
CAT#438 Deluxe Quarto Hardcover Edition. Black Suede w/Gilt Lettering. $395.00

W. D. GANN'S MASS PRESSURE FORECASTING CHARTS
Daniel T. Ferrera

2004. 103p. This new book by Daniel T. Ferrera develops a theory of how Gann most probably created his Mass Pressure Forecasting Charts. "Of all the Gann forecasting techniques known, the Mass Pressure Formula has been one of the most closely guarded secrets. In fact, there are very few "Gann Experts" that even know how to create a Mass Pressure Chart or anything about the nature of its construction. These charts are based entirely on Gann's philosophy that "the future is nothing but a repetition of the past". Each year, W.D. Gann would draw up a stick figure forecast of the stock market averages and various commodity futures in his Supply & Demand newsletter service. He would provide commentary that would provide an outline of what he expected the market to do. Gann said that the future was just a repetition of the past and one needed only the right beginning to predict the future. In later years, Gann provided something he called the Mass Pressure Chart. In his advertisement for his Master Mathematical Time Price and Trend Calculator, Gann said that he would provide a Mass Pressure Chart for the upcoming year. Based upon its vague description, the Mass Pressure Chart is supposed to indicate bullish and bearish trends according to Gann's Master Time Factor. I have talked to many Gann experts about the Mass Pressure Chart and most have no clue as to what it may be. I have never seen an actual original Mass Pressure Chart from W.D. Gann and I do not know of anyone else who has come across one. This course presents what I feel is most likely the solution to this question." Dan Ferrera. Contents: Instructions for Using the Excel Mass Pressure Worksheet. Mass Pressure Forecasting Article from Traders World Magazine. W.D. Gann's Mass Pressure Chart. W.D. Gann's Secrets to Forecasting. Forecasting Monthly Moves. W.D. Gann's 1929 Forecast Recreated. Special Stock Market Cycle Report I (Free Bonus). Special Stock Market Cycle Report I (Free Bonus). W.D. Gann's Original Forecasting Course. Includes CD ROM with Detailed Excel Spreadsheets for Creating Mass Pressure Charts, & Spreadsheet Duplicating Gann's 1929 Forecast.
CAT#495 Deluxe Quarto Hardcover Edition. Black Suede w/Gilt Lettering. W/CD ROM $195.00

STOCK MARKET PREDICTION: *THE HISTORICAL & FUTURE SIDEROGRAPH CHARTS &*
SOFTWARE. Donald Bradley. (Charts & Software by Daniel Ferrera)

1948, 2004 180p. The Planetary Barometer & How to Use It. Donald Bradley's Siderograph Indicator is a very popular market indicator used by many analysts to give current turning points and trend indications for the markets. We have kept Bradley's original work available for many years, but there has never been software to produce the charts readily available. Upon learning that services charged up to $100 per year for each Bradley chart, and that the only available software to produce the charts ran into the thousands of dollars, we put together a new version of Bradley's work including the Bradley Siderograph charts for 100 years from 1950 to 2050, including the software so that anyone can produce these charts for themselves. The historical charts will also allow researchers to track past performance of the Siderograph

for the past 55 years of market history, and to have ready-made charts covering the next 45 years. This book includes a CD ROM with the software, and is bound in a Hardcover. CONTENTS: Mystery of Mass Psychology; Human Response to Outside Forces; Tides In Affairs of Men; Cycles Write World History; Search for Causes; 3 1/2 Year Business Cycle; Planetary Aspects Are The Secret; Jupiter-Uranus Cycle; Planetary Periods & Synods; Symbols; Aspects in Action; Power of Aspects; Line of Aspectivity; computing; Example; Sideograph; Promises & Limitations. 100 Years of Siderograph Charts.
CAT#496 HARDCOVER W/CD ROM. PRICE $195.00

FERRERA COMPLETE OUTLOOKS – 2008 – 2015
(INDIVIDUAL OUTLOOKS FOR EACH YEAR ALSO AVAILABLE ANNUALLY)
Daniel T. Ferrera

Sep, 2008 – Dec, 2015. Dan began these yearly Outlooks in 2008, and has continued the tradition for 8 years now. His sequence of yearly Outlooks serves as more than just a financial road map, they present a detailed course on applied technical analysis and forecasting by one of the leaders in the modern field. The greatest insight is gained by studying the entire sequence and logic of the Ferrera Outlooks from 2008 progressively through 2015. After studying these, you will never approach a market in the same way! For new customers who have not subscribed in the past, we offer this special package of the complete Outlooks from 2008-2015 at a discounted price!
PDF FILES DELIVERED BY EMAIL CAT#237 PRICE $500.00

STUDIES IN ASTROLOGICAL BIBLE INTERPRETATION
Daniel T. Ferrera

2001. An interesting exploration of the process used in coding astrological and astronomical cycles into literature. Engages in a thorough analysis of the book of Genesis, exploring coding systems by which astrological symbolism is veiled. Contents: Study of George Bayer's Bible Interpretation; A Study of Ludwig Larson's Key to the Bible & Heaven; A Study of David Fideler's Jesus Christ Sun of God; Revelations Revisited; Bible Interpretation Related to W. D. Gann; 666 The Number of the Beast; A Study of the Book of Genesis; The Number 12; The Great Flood; Astrological Analysis of Astrological Codes of Genesis; Noah & His Sons; The Complete Book of Genesis is Broken Down Into Astrological Symbolism. Includes two horoscopes of the Bible.
CAT#440 Deluxe Quarto Hardcover Edition. Black Suede w/Guilt Lettering. $75.00

GANN FOR THE ACTIVE TRADER
New Methods for Today's Markets
Daniel T. Ferrera

2006. 150p. In this new book, Gann expert Dan Ferrera presents a number of new techniques for trading in today's markets which build on and expand the trading methods of the legendary trader of yesteryear, W.D. Gann. It is exceptionally difficult to learn how to use Gann's methods effectively…and this outstanding new book is a treasure chest for those interested in Gann's work. Includes a bonus 80 page Gann mini-course! This book serves as an excellent introduction to the work of Dan Ferrera giving some introductory techniques using principles from the work of W. D. Gann. For those who would like to review his work and through process before purchasing his more specialized and advanced courses, this work is a cheap and easy place to begin. Contents: Trading is a Profession; How Much Do You Need to Start Correctly; Commodity Basics; Risk Disclosure; The Most Neglected Trading Discipline; W. D. Gann's Most Important Money Management Rules; Understanding the Basics; Understand Trends and Trend Line Breaks; Market Swings; Support and Resistance; Short Term Consolidation Patterns; Trends Again: Bar Grouping Technique; Swing Trading: A Quick Word about W.D. Gann; Explaining Gann's 50% Rule; Gann's Red Light Green Light Indicator; ABC's & 123's; Putting it all Together; Using Inside Bars to Enter in the Direction of the Larger Trend; Understanding the Options Opportunity; Timing Important Stock Market Bottoms; Successful Trading; What is Luck?; Bonus Material: W.D. Gann Mini-Course.
CAT#517 Deluxe Quarto Hardcover Edition. $75.00

Cosmological Economics

The Masters Of Technical Analysis Series

The Masters of Technical Analysis Series brings together a collection of the most important classical and modern works on technical analysis and financial market forecasting. These classic works from the Golden Age of Technical Analysis were carefully selected by the late Dr. Jerome Baumring of the Investment Centre Bookstore in the 1980's, as representing the most valuable and important works in technical analysis ever written. They were included as the foundational source texts for his program in advanced financial market analysis and forecasting, and serve as the ideal foundation for any analyst seeking a thorough education in market theory and technical trading.

The Golden Age of technical analysis was a period from the early 1900's through the 1960's where the foundational theories of modern financial analysis were initially developed. The ideas and technologies developed during this fruitful period have formed the basis for most modern technical market theory, which is considered to be mostly a repetition or reworking of these past ideas and techniques developed by the Old Masters of the Golden Age. In these historical works can be found the timeless trading wisdom which has laid the foundation for all modern investment theory and literature. These techniques are as useful in today's markets as they were in the past, providing rare and valuable insights, tools and strategies that give the modern trader an edge over traders and investors that are unaware of these time honored tools.

Each quality reprint of these classical texts has been reproduced as an exact facsimile of the original text, maintaining the original layout, typeset, charts, and style of the author and time period, helping to preserve and communicate a sense of the feeling of the original work that a reproduction in modern format does not capture. Many of these rare works and courses were originally printed in only very small private editions or as correspondence courses, so that the originals were easily lost or destroyed over time. Our reproductions of these important source works have been printed on acid free paper and bound in a quality hardcover format that will compliment any trading library and help to preserve this important resource for generations to come.

The series is also currently being digitized and archived for permanent digital preservation by the Institute of Cosmological Economics, creating a searchable reference library of market wisdom accessible globally and available in new digital formats to keep the knowledge fresh and accessible through new devices and technology as we advance further into the information revolution. To see our full catalog of hardcover reprints, new publications, and digital editions please visit our website at www.CosmoEconomics.com.

- ❖ **Samuel Benner, An Ohio Farmer** - Benner's Prophecies - *Of Future Ups and Downs in Prices* - (1879)
- ❖ **Geo. W. Cole, La Marquette** - Graphs & Their Application to Speculation - (1936)
- ❖ **Frank Tubbs** - Tubbs' Stock Market Correspondence Course - (1944)
- ❖ **R. N. Elliot** - Collected Works of R. N. Elliot - *The Wave Principle. Nature`s Law: The Secret of the Universe. "The Wave Principle": A Series of Articles Published in 1939.* - (1946)
- ❖ **Joseph A. Wyler** - Wyler Series on Stock Market Speculation - *Vol.1, The Application Of Scientific Principles To Stock Speculation. Vol.2, Trading And Trending* - (1960)
- ❖ **Edward Dewey** - How to Make a Cycles Analysis - *Correspondence Course in Advanced Cycle Analysis*- (1955)
- ❖ **Nikolai D. Kondratieff** - Long Waves In Economic Life - (1935)
- ❖ **Pickell & Daniel** - Pickell-Daniel Extension Course of Market Analysis - (1937)
- ❖ **Franklin Paul Jackson** - The I-S Method - *Individual Stock - Intermediate Swing* - (1972)
- ❖ **Richard Schabacker** - Technical Analysis & Stock Market Profits - (1930)
- ❖ **Richard Schabacker** - Stock Market Profits - (1934)
- ❖ **M. V. Woods** - Seven Studies In Stock Market Trading - (1943)
- ❖ **William Dunnigan** - Collected Works of William Dunnigan - *Gains in Grains. New Blueprints for Gains in Stocks & Grains; Barometers for Forecasting Stocks. One-Way Formula for Trading In Stocks & Commodities* - (1957)
- ❖ **Henry Ludwell Moore** - Collected Works of Henry Ludwell Moore - *Economic Cycles Their Law & Cause. Generating Economic Cycles. Forecasting The Yield of Cotton* - (1923)
- ❖ **Edwin S. Quinn** - Action - Reaction Signals - (1950)
- ❖ **Emil Schultheis** - Basic Trend Barometer - *A Long Term Stock Trend Study* - (1946)
- ❖ **Payson Todd** - The "Todd Theory" of Market Measurement & Price Projection - (1953)
- ❖ **Dr. Alexander Goulden** - Behind The Veil - (2010)
- ❖ **C. S. Johnson, C. P. A.** - A New Technique of Stock Market Forecasting - (1931)
- ❖ **William D. Gann** - The Collected Writings of W. D. Gann, Volume I - *Marketing Brochures, Interviews, Annual Forecasts & Trading Records* - (1909-1954)
- ❖ **William D. Gann** - The Collected Writings of W. D. Gann, Volume II - *The Master Time Factor: No. 3 Master Forecasting Method & Stock Market Forecasting Courses* - (1921-1954)
- ❖ **William D. Gann** - Collected Writings of W. D. Gann, Volume III - *Master Mathematical Formula & Calculators* - (1955)
- ❖ **William D. Gann** - Collected Writings of W. D. Gann, Volume IV - *The Complete Commodity Courses* - (1940)
- ❖ **William D. Gann** - Collected Writings of W. D. Gann, Volume V - *Introductory Stock Market Courses, Mechanical Methods, & Trend Indicators Courses* - (1935-1950)
- ❖ **William D. Gann** - Complete Collected Writings of W. D. Gann - In 6 Volumes - (1955)
- ❖ **Timothy Walker** - How To Trade Like W. D. Gann - *An Exploration of the Mechanical Trading Lesson on U. S. Steel* - (2014)
- ❖ **James P. Morton** - When to Sell to Assure Profits - (1926)
- ❖ **Daniele Prandelli** - The Polarity Factor System - *An Integrated Forecasting & Trading Strategy Inspired By W. D. Gann's Master Time Factor* - (2012)
- ❖ **Daniele Prandelli** - The Law Of Cause And Effect - *Creating A Planetary Price/Time Map Of Market Action Through Sympathetic Resonance* - (2010)
- ❖ **Perspectives, National Graphic Co.** - The Great Bull Market & Collapse - (1932)
- ❖ **Richard Martin** - An Introduction to Trend - Action - *A Scientific Method of Forecasting* - (1943)

- ❖ **George Bayer** - Money Investing In Stocks, Trading In Commodities, Or The Time Factors In The Stock Market - *The Art of Scientifically Detecting Direction & Distance of Swings* - (1937)
- ❖ **George Bayer** - The Egg of Columbus - (1942)
- ❖ **George Bayer** - Stock & Commodity Traders Hand-Book of Trend Determination - *Secrets of Forecasting Values, Especially Commodities, Including Stocks* - (1940)
- ❖ **George Bayer** - Gold Nuggets for Stock & Commodity Traders - (1941)
- ❖ **George Bayer** - Preview of Markets - *VOL. I, NOS. 1-10* - (1939)
- ❖ **George Bayer** - George Wollsten - Expert Stock & Grain Trader - (1946)
- ❖ **George Bayer** - The Collected Works of George Bayer - *9 Books In 2 Hardcover Volumes* - (1939)
- ❖ **Cliff Stewart** - Magic of Making Money in The Stock Market - (1951)
- ❖ **Dickson F. Watts** - Speculation as a Fine Art - *& Thoughts On Life* - (1865)
- ❖ **Henry Ansley** - I Like the Depression - (1932)
- ❖ **Henry Hall** - How Money is Made in Security Investment - *Or A Fortune At Fifty-Five* - (1908)
- ❖ **Dr. Jerome Baumring & Julius J. Nirenstein** - Gann Harmony: The Law of Vibration: The Complete Course Manuals & Lecture Notes - *The Complete Gann 1-9 Course Manuals. Compiled By Dr. Jerome Baumring With Notes On W. D. Gann's Hidden Material: The Complete Gann 1-9 Lecture Notes* - (1989)
- ❖ **Dr. Jerome Baumring** - Gann Harmony: The Law of Vibration. The Complete Course Manuals - *Course Manuals For Gann 1 Through Gann 9* - (1989)
- ❖ **Julius J. Nirenstein & Dr. Jerome Baumring** - Notes on W. D. Gann's Hidden Material: The Complete Lecture Notes - *Lecture Notes for Gann 1 through Gann 9* - (1989)
- ❖ **Alfred Friedman** - Lecture Notes from Baumring's Investment Centre Series - (1987)
- ❖ **B. Edlin** - Minor Swings of the Stock Market and their indications - (1924)
- ❖ **Henry Howard Harper** - The Psychology of Speculation - *The Human Element In Market Transactions* - (1926)
- ❖ **S. A. Nelson** - The ABC of Options & Arbitrage - (1904)
- ❖ **Daniel T. Ferrera** - The Gann Pyramid - *Square of Nine Essentials* - (2001)
- ❖ **Daniel T. Ferrera** - The Mysteries of Gann Analysis Unveiled! - *A Detailed Presentation of W. D. Gann's Technical Trading Principles* - (2001)
- ❖ **Daniel T. Ferrera** - Wheels Within Wheels – *The Art of Forecasting Financial Market Cycles* - (2002)
- ❖ **Daniel T. Ferrera** - W. D. Gann's Mass Pressure Forecasting Charts - (2004)
- ❖ **Daniel T. Ferrera** - The Keys to Successful Speculation - (2004)
- ❖ **Daniel T. Ferrera** - Spirals of Growth and Decay - *Exposing the Underlying Structure of Financial Markets* (2005)
- ❖ **Daniel T. Ferrera** – Gann for the Active Trader - *New Methods For Today's Markets* (2006)
- ❖ **Daniel T Ferrera** – Economic & Stock Market Forecasting – *W.D. Gann's Science of Cyclical Periodicity Sequencing* (2013)
- ❖ **Daniel T. Ferrera** - The Path of Least Resistance - *The Underlying Wisdom & Philosophy of W. D. Gann Elegantly Encoded in the Master Charts* - (2014)
- ❖ **Warren Hickernell** - What Makes Stock Market Prices - (1932)
- ❖ **William C. Moore (Market Expert)** - Wall Street - *Its Mysteries Revealed, Its Secrets Exposed.* - (1932)

www.ingramcontent.com/pod-product-compliance
Lightning Source LLC
Chambersburg PA
CBHW070244230326
41458CB00100B/6076